Birds of Delhi

Birds of Delhi

RANJIT LAL

UNIVERSITY PRESS

YMCA Library Building, Jai Singh Road, New Delhi 110 001

Oxford University Press is a department of the University of Oxford. It
furthers the University's objective of excellence in research, scholarship, and
education by publishing worldwide in

Oxford   New York

Auckland   Bangkok   Buenos Aires   Cape Town
Chennai   Dar es Salaam   Delhi   Hong Kong   Istanbul   Karachi   Kolkata
Kuala Lumpur   Madrid   Melbourne   Mexico City   Mumbai   Nairobi   Sao
Paulo   Shanghai   Taipei   Tokyo   Toronto

Oxford is a registered trade mark of Oxford University Press
in the UK and in certain other countries

Published in India
by Oxford University Press, New Delhi

© Oxford University Press 2003

The moral rights of the author have been asserted

Database right Oxford University Press (maker)

First published 2003
Oxford India Paperbacks 2004

All rights reserved. No part of this publication may be reproduced,
stored in a retrieval system, or transmitted, in any form or by any means,
without the prior permission in writing of Oxford University Press, or as
expressly permitted by law, or under terms agreed with the appropriate
reprographics rights organization. Enquiries concerning reproduction outside
the scope of the above should be sent to the Rights Department, Oxford
University Press, at the address above

You must not circulate this book in any other binding or cover and you must
impose this same condition on any acquirer

ISBN 019 5672194

Typeset by Wordsmiths, Delhi 110034
Printed by Karan Press, Delhi 110020
Published by Manzar Khan, Oxford University Press
YMCA Library Building, Jai Singh Road, New Delhi 110 001

Author's Note

The landscape of Delhi seems to be changing faster than ever. In just a few months, new roads, new flyovers and the Metro rail project have rendered some areas unrecognizable, and sadly, most of this means bad news for the birds of the city. Urban development along the banks of the Jamuna is playing havoc with waterside habitats, and grand old trees that provided safe nesting for scores of birds of many species are being felled indiscriminately since their inability to move makes them a major obstruction in the execution of the Metro project. Alipur Road in north Delhi has, for example, suffered hugely; the skyline here has changed entirely, and the poor remnants of the Qudsia gardens are now under attack. On the plus side, however, birdwatching is rapidly gaining popularity amongst the capital's citizens with the re-activated Delhi Bird Club spearheading the way. One of its leading lights, Bill Harvey, recently discovered a wonderful water-bird habitat at Basai, just outside Gurgaon and *en route* to Sultanpur, by watching a flock of bar-headed geese land while he stood waiting at a railway crossing! As a water-bird habitat, Basai is now rivaling Sultanpur for supremacy. Let us hope there are many more Basais waiting to be discovered in the areas around Delhi.

R. L.

June 2002
Delhi

Preface

In this book you will meet several of Delhi's more colourful, interesting, and remarkable inhabitants—its birds. Any newcomer to the city will have become aware (even if dimly so) of some of its avian inhabitants—the gurgling rock doves that are fed at street corners, sparrows that shrilly strut around like roadside Romeos, 'green chillie' parakeets tearing apart the evening skies with their screams, *myna*s engaged in raucous brawls in parks, crows clearing up after messy picnic parties, and kites diving down from the skies to pick up run-over rats from the streets.

There are, however, many more, of bewildering variety, colour, and habits. As we go along, you will learn where they hang out, when (and whether) they come and go, and where and when they nest. You will also learn of the dangers they face and their uncertain future as we heedlessly continue to encroach upon and destroy their habitats. While I have mentioned specific sites where I have seen particular species, these are by no means all-inclusive; these only typify the habitat the bird prefers; thus any park or large garden with great, leafy, ancient trees is likely to harbour owls, hornbills, parakeets, mynas, barbets, and other hole nesters. The sites I have mentioned are those I am familiar with.

The order in which the birds appear is based on the relatively recent (1988) Sibley and Munroe system of classification, which links bird families together by the closeness of their DNA affiliations. I have also used the recently introduced common names from *An Annotated Checklist of the Birds of the Oriental Region* by Inskipp et al.,[1] but have, for convenience, also mentioned the 'old', often more familiar, names within brackets. Names in Hindi follow the scientific name in order of appearance. Thus, for example: Grey Francolin (Grey Partridge), *Francolinus pondicerianus*, Teetar.

I have, in this book, quite deliberately emphasized on the 'here' and 'now' of the birds of Delhi, and touched but lightly upon the abundance of the past. Those of you interested in bygone birding may see Usha Ganguli's wonderful classic, *A Guide to the Birds of the Delhi Area* and H. P. W. Hutson's *Birds about Delhi*,[2] though both are now out of print and may be difficult to get hold of.

Delhiites are gradually becoming aware of, and enthused by the wonderful variety of birds that share their city with them. With this book, I hope that many many more are encouraged to do so.

---

[1] Inskipp T., N. Lindsey and W. Duckworh, *An Annotated Checklist of the Birds of the Oriental Region*, Oriental Bird Club: Sandy, UK, 1996.

[2] Usha Ganguli, *A Guide to the Birds of the Delhi Area*, Indian Council for Agricultural Research, 1975; H. P. W. Hutson, *Birds about Delhi*, Delhi Birdwatching Society, 1954.

Contents

	Preface	vii
	List of Plates	x
I	THE FLIGHT OF THE SHIKRA	1
	A Bird's-eye View of Delhi	3
	Some Popular Birding Areas	14
	Open Skies, Open Seasons	23
	Fields Notes from Three Typical Birding Trips	29
II	THE ROGUES' GALLERY	35
	The Birds	37
	Index of Birds Featured in the Rogues' Gallery	147

List of Plates

(Between pp. 20–1)

Plate nos

1. White-rumped Vulture
2. Shikra
3. Greater Flamingos
4. Spotted Owlet
5. Painted Stork and family
6. Spotbill Duck
7. Magpie Robin
8. Ruff

(Between pp. 52–3)

9. Gulls
10. Grey Francolin
11. Black Francolin
12. Northern Pintail
13. Indian Grey Hornbill
14. Common Hoopoe
15. Indian Roller or Blue Jay

List of Plates ◆ xi

16.	White-throated Kingfisher
17.	Pied Kingfisher
18.	Small Green Bee-eater
19.	Blue-checked Bee-eater
20.	Great-white Pelicans
21.	Plum-headed Parakeet
22.	Barn Owl
23.	Laughing Dove
24.	Yellow-footed Green Pigeon
25.	Common Moorhen
26.	Common (or Fantail) Snipe
27.	Blackwinged Stilt

(Between pp. 68–9)

28.	Red-wattled Lapwing
29.	Darter
30.	Indian Cormorant
31.	Egrets

(Between pp. 100–1)

32.	Pond Heron
33.	Purple Heron
34.	Little Heron
35.	Black-necked Stork
36.	Bay-backed Shrike
37.	Grey Shrike
38.	Rufous Tree-pie
39.	Small Minivet
40.	Long-tailed Minivet

xii ◆ *List of Plates*

41.	Black Drongo
42.	Asian Paradise Flycatcher (Male)
43.	Asian Paradise Flycatcher (Female)
44.	Black Redstart
45.	Common Stonechat
46.	Brahminy Starling
47.	Asian Pied Starling
48.	Wire-tailed Swallow
49.	Red-whiskered Bulbul
50.	Red-vented Bulbul
51.	Ashy Prinia
52.	Jungle Babbler
53.	White-browed Wagtail
54.	Red Avadavat or Red Munia
55.	Silverbill or White-throated Munia

I
THE FLIGHT OF THE SHIKRA

A Bird's-eye View of Delhi

A comprehensive checklist of birds brought out in the early 1990s by the environment non-governmental organization (NGO) Kalpavriksh, based on its own findings and other observations, listed a grand total of nearly 450 species for Delhi and its surrounding areas (i.e. within a 50 km radius) recorded in the last two hundred years or so. This is around one-third the total number of species recorded in the country. While this impressive tally would have certainly declined since, Delhi still attracts a kaleidoscopic variety and enormous number. Just what is it then that draws birds so compellingly to the capital?

Most significant perhaps is the capital's interesting and diverse mix of habitats and 'ecosystems'. The most prominent physical feature of the capital is, of course, the Ridge, which is actually the tail end of the Mewat branch of the ancient Aravalli mountains. The Ridge marches into Delhi from the south, from Gurgaon, in the form of a plateau some 5 km across. Before entering the city proper, it splits into two ranges: one turns sharply to re-enter Gurgaon, and the other, which continues in a north-easterly direction, virtually bisects the city before tapering off at the west bank of the Jamuna.

The southern range of the Ridge retains much of its original badland character—rocky, with hardy thorn scrub and tough semi-desert flora, and is home to great numbers of francolin (partridge), peafowl, quail, rock chats, owls, wren-warblers, and raptors of various kinds. Much of this sparawling area has, however, now been built up or ravaged by quarrying.

The other branch of the Ridge, comprising the New Delhi or central Ridge (864 ha), and the northern or Kamala Nehru Ridge (87 ha) forms the vital green lungs of Delhi. Artificially afforested by the British at the beginning of the last century, the dense canopy of acacias (of several kinds), *neem*, *palas*, and numerous other species of trees, provides home and shelter to around two hundred species of birds.

The Jamuna, its backwaters and lagoons (formed by the Okhla Barrage), the low-lying land surrounding it (locally known as *khader* and subject to regular flooding), as well as the fields adjoining its *bund*s, constitute Delhi's other major 'ecosystem' and bird habitat. Flowing south from Wazirabad to Okhla, the river and its sand banks and extensive reed beds are host to an enormous variety of wildfowl, waders, and passerines, both migratory and resident.

Apart from these natural and naturalized ecosystems, there are man-made habitats that have their own attractions. Several of the city's parks have impressive stands of fruiting and flowering trees, which attract a host of birds. Birds have long used the city's historical monuments and buildings for nesting and roosting.

While there are a large number of tough, hard-core species that either live (and breed) in the capital the year round, or come here simply to breed in summer, a slightly higher number of species arrive here in winter from the Himalayas or further afield—eastern Europe, Central Asia, and even Siberia. The capital is well located for migratory birds funnelling through the river valleys of the Indus and Brahmaptura. Some of these migrants spend the entire winter in the city, others pass through, stopping awhile en route to destinations further south, or while on their way back home north, in spring.

Diversity in climate and habitats has combined to give Delhi a range of bird-life perhaps unmatched by any other metropolitan city in India. Let us take a brief flight over the city now, for a proper bird's-eye view of the capital...

The Flight of the Shikra

High above the stolid sandstone memorial of India Gate, a shikra flutters in small, tight circles in a hot blue sky. It is not hunting just yet—this fierce diminutive hawk with its burning golden eyes and fine bluish-grey plumage—only enjoying itself and scanning the scene below. Sprawled out beneath, stretching to every horizon is Delhi, with its ancient history, graceful monuments, cacophonous traffic jams, great rocky woodlands, and sweeping swards of green—home, refuge, winter retreat, and stopover point for nearly 450 species of birds that have been recorded here during the last 200 years or so. Today, of course, there are far fewer; if you begin birdwatching in earnest you'd be doing well to tick off around 200–50 species (approximately half of which will be migrants) in your first two or three years. Thereafter, each new species will be harder to come by and something of a bonus; but these delightful surprises will keep cropping up.

If the shikra were to fly due north from India Gate, it would soon pass over the clamorous walled city of Shahjehanabad (with its avid pigeon fanciers) and thereon, over the historical Qudsia Gardens, where barn owls and grey hornbills have nested for many, many generations in its gnarled ancient trees. Still heading north, the bird would soon be flying over the great Jamuna, with its high bunds lined with eucalyptus and acacias out of which bee-eaters and drongos sally forth. The cultivated sandy riverbed is shrill with stilts and lapwings, and guarded warily by pond herons and egrets. If the shikra followed the turbid waters southwards now, past the Inter State Bus Terminal and clanging Railway Bridge right down to the peaceful, verdant lawns of Rajghat, it would fly over what, till as late as the 1980s, was a very rewarding area for birding. On a winter morning, along this 8-km stretch of riverbank and surrounding parks and fields, you could see almost 100 species quite easily. Bridge building, the Metro Rail project, and just too many people and vehicles swarming about the riverbank have destroyed much of this once rich bird habitat.

Continuing its way south along the river, the shikra would have to dodge the fearful Indraprastha Power Station and then turn west to fly over the burly Purana Quila (Old Fort) and the National Zoological Park—which has always been more attractive as a haven for wild birds than as a zoo. Here, colonies of painted storks, black-crowned night herons, egrets, and cormorants have nested in comfortable chaos for generations, and in winter, migratory waterfowl like pintail, shoveller, and common teal paddle the ponds with city-bred savoir-faire. South of the zoo lies Humayun's tomb in a garden of great trees where orioles gleam during the rains and house swifts flicker and squeak around the graceful domes, diving unerringly into the dark doorways. East of this, the hawk is back over the river and following its course southwards again, soon flying over packed conglomerations of water hyacinth where bitterns and herons stand turned to stone, and snazzy pheasant-tailed jacanas stride elegantly over the waxy leaves on giant spider-feet. Here, as one approaches Okhla Barrage, the river, turgid and filthy as it is, teems with bird-life. Tall, statuesque grey herons man the mudbanks at regular intervals while plovers and sandpipers scurry and dart in the mud, like clockwork toys. Pond herons, streaked in dry grass plumage, stand frozen in ambush, before startling you by unfolding dazzling, white wings. A large flock of flamingos, with their pink-gin tutus and ballerina poses, add a tinge of glamour to the pewter-coloured waters, and a platoon of snazzy pied avocets go about minesweeping in the ooze with great verve and industry. Skeins of shell-white egrets fly languidly by on translucent wings, while squadrons of witch-black cormorants are as purposeful and intent as fighters on a search-and-destroy mission. Terns, with silver-blade wings laconically patrol up and down, their black-capped heads riveted downwards on the water. River lapwings strike majestically Napoleonic poses as they try to intimidate one another, and the skewer-billed, black-winged stilts stand tall on pink knitting-needle legs busy rolling in the muck everywhere. By October, huge flocks of duck and coot freckle the water while the great Pallas gull with its smoky charred-wood and white plumage, wings majestically past, giving them the critical once-over like a sergeant-major on drill inspection.

This is an ever-changing waterscape. When the Barrage holds the river back, a large, calm lake is formed behind its sluice gates; a haven for wild fowl and water-birds. If the water level falls, or the lake is drained out (usually depending on agricultural needs), large tracts of muddy marsh are exposed to which waders flock excitedly—this site is thus best visited with an open mind, for you can never really know what to expect.

The shikra would, however, be more interested in the silvery plumes of *moonj* grass shooting up 10 feet high in the fields and khader wastelands beside the river south of the Barrage. This is *munia* and warbler country, as well as a good place to look out for francolin (partridge) and quail: fine hunting grounds for a hawk. Yellow-eyed babblers cling to reed stems and trill nervously, glaring at you out of red-rimmed eyes. Larks spring skywards, pouring forth musical-box melodies and elfin green bee-eaters skate the breezes skilfully, snapping up dragonflies with élan. Great, lumpy crow pheasants sun their russet cloaks and whoop ghoulishly as they flap clumsily from bush top to bush top in their search of baby birds. During the monsoons, colonies of *baya*s—which look like sparrows with fog-lamp heads and chocolate-chip markings—weave graceful, vase-like homes in the spiky date palms while orioles, doves, mynas, and drongos nest companionably together in the glittering, scimitar-leaved neems. Demure pied bushchats keep watch from bush tops in pairs; the male in black and white, his wife, biscuit brown. The shrikes, with their 'masks of Zorro' and executioner bills keep a sharp eye out from the tops of bushes and posts, for the incautious frog or lizard. And everywhere, dun and anonymous warblers leap and flit excitedly from grass tuft to grass tuft and bush to bush shouting their heads off.

But now it's time to turn south-west again, towards the forbidding granite fortress of Tughlakabad. Within its burly walls, lies a rocky, ankle-twisting landscape, furnished with thorn bushes and scrub. Here our shikra will have competition, for this is good raptor country with buzzards and hawks manning the massive battlements—as though on the lookout for invaders. Once, these ramparts were great take-off spurs for vultures that all but disappeared from Delhi's skies but now seem to be making a comeback. This is one of the few places where you can still meet the quaint and increasingly rare, yellow wattled lapwing looking comically avant-garde with it canary-yellow plastic moustache and matching legs. A short flight south from the fort lies the recently established Asola wildlife sanctuary, ringing with the calls of black and grey francolin, its bush tops policed by the ever-snazzy shrikes. Indian robins, bush-larks, wren-warblers and munias abound in this acacia-clad, raw rock landscape and bands of red-vented bulbuls and bank mynas do their rounds with pugnacious efficacy.

Heading north-west from this tough, big-boulder country, the hawk is soon flying over the cheek-by-jowl 'posh' colonies of south Delhi. Fortunately, nearly every one of them has a green area or park within its neighbourhood: Thus there is the rambling Hauz Khas woodland and adjacent Deer Park near Hauz Khas and Green Park (where 100 bird species have been recorded), the Jahapanah City Forest off Greater Kailash II (boasting 70 species), and, further north, the celebrity-status Lodi Gardens in central Delhi. In these, you can acquaint yourself with the common woodland birds of Delhi: its resident babblers, barbets, woodpeckers, *bulbul*s, owls, parakeets, hornbills, mynas, tailorbirds, bee-eaters, magpie robins, kingfishers, et al. In winter, when the migrants arrive, there are redstarts, wagtails, flycatchers, and lesser whitethroats to add to the tally, as well as beauties like the rose-finch and pop-star-looking, well-gelled, spangled drongo.

Just south of these colonies sprawls more of the rocky Ridge country, where the Qutub Minar stands tall and rock chats, owls, buzzards, drongos, buntings, and Indian robins make themselves at home. The famous and deep, 1 km long 'canyon' in the sprawling Jawaharlal Nehru University campus, has for years been the haunt of the great horned owl, and its surrounding acacia-clad wilderness quivers with partridge, peafowl, babblers, and bulbuls. Heading north from here, towards the city, the shikra will soon find itself over that great emerald lung of Delhi, the central or New Delhi Ridge. This vast, woolly pelt of green virtually splits the city diagonally, and along with its flinthead-shaped northern spur (the northern or Kamala Nehru Ridge) harbours around two hundred species of birds. Here you may see the paradise flycatcher flutter its satiny ribbons entrancingly through the foliage, or the rare blue-headed rock thrush gleaming sapphire and amber in a smoky beam of sunlight. This is good raptor country too and scanning the skies here may reveal black-winged kites, buzzards, eagles, and shikras. In the parks that abut the Ridge, like Buddha Jayanti Park for instance, you can see parakeets, munias, weavers, doves, blue rock pigeons and even rose-finches (in winter) partake of a breakfast of birdseed, bread, and dried corn seed laid out for them (a practice common in all parks in the city). Due east of the New Delhi Ridge lie Rashtrapati Bhavan and the President's Estate, and beyond that Rajpath takes us straight back to India Gate from where the journey began. Several of New Delhi's famous, tree-lined avenues radiate out of here, the trees being home, or providing shelter, to hornbills, owls, barbets, magpie robins, mynas, and a host of others. Egrets and kingfishers mark time along the long straight water troughs and redwattled lapwings patrol the lawns of India Gate with bureaucratic officiousness.

Two storm-water drains wind their way through Delhi and support rich bird-life too. There is the Najafgarh Nallah in west Delhi, which is good for water-birds, but which, unfortunately, turns into a dead sewer after entering the industrial belt. There is also Khushak Nallah, originating in the Ridge and flowing into the Jamuna, through some dense scrub jungle teeming with birds.

So much for public places. Private gardens and lawns in Delhi (and the 'small holdings' of Delhi's millionaire farmers at Mehrauli) provide sanctuary too, in no mean measure. Even a small garden with fruiting or flowering trees and bushes may attract perhaps fifty species of birds during the course of a calendar year. Built-up areas too are not ignored; black kites dive and swirl through the strangling maze of cables and wires in the walled city, to pick up offal from the shuddering streets. Blue rock pigeons nest brainlessly, even in the service bays of garages, and peafowl, bereft of tangled undergrowth, may chose a quiet porch or terrace, and lay their clutch of eggs there. Every tangled bush of bougainvillaea and *madhumati* (the Rangoon creeper) will have doves, bulbuls, warblers, and babblers nesting within.

Several rich bird habitats lie within a couple of hours' drive of Delhi. One of the better known of these is Sultanpur National Park, 15 km beyond Gurgaon, in neighbouring Haryana. This tiny 1.5 sq km seasonal wetland attracted nearly 250 species of birds during its heyday in the 1980s and early 1990s. But then the lake dried up, due mainly to man-made causes, and the birds deserted it *en masse*. Water from a canal is now being supplied to Sultanpur and it seems that the birds have welcomed this development. It may take a little time before the ecosystem stabilizes again, but hopefully Sultanpur will eventually regain some of its lost species and glories.

Other water bodies around the capital include Badkhal Lake, adjacent to Faridabad, a deep-water body where you can see red crested (and other) pochards in winter. Bhindawas bird sanctuary in the Rohtak district of Haryana, was another splendid water-bird habitat, but in recent years has all but been overwhelmed by water hyacinth.

Near Sonepat, in Haryana, passing under National Highway No. 1 (the notorious Grand Trunk Road), winds another stormwater *nallah*, unimaginatively called Drain No. 8. It is, however, immensely rich in bird-life and more than 160 species have been sighted along its banks.

How, then, have the people of Delhi got along with their avian companions? In the days gone by, and to a much lesser extent even now, birds—especially waterfowl, waders, doves, and partridges— were hunted mercilessly. Today, owls, parakeets, and munias are still persecuted, the first, because their body parts are used in dark occult practices, and the latter two thanks to the pet trade. Most Delhites, however, seem blindly oblivious to the dazzling wealth of bird-life around them. Happily, there are a large number of people who regularly put out birdseed, bread crumbs, and water for them every morning in the city's parks and gardens, though few stay on to watch the proceedings. In fact, in many such places the birds have come to expect breakfast on the table every morning, and I have often seen the black-rumped flameback (the ex-golden backed woodpecker), hitch itself down a tree trunk and hop sideways on to one such feeding spot, to pick up (with a horizontal beak!) the bread crumbs that had been scattered there.

As for birdwatchers or birders, there have long been small but enthusiastic groups of people who are out every morning with binoculars and scopes, ticking off species, making lists, and following the flight of the shikra.

The first birding groups consisted chiefly of expatriates whose observations, notes, and lists have formed the basis of our knowledge of Delhi's rich bird-life. In the mid-1920s, Sir Basil-Edwardes collected specimens of 230 species within a five-month period. Sir N.F. Frome listed some three hundred species in 1947, based on observations made by him and others between 1931 and 1945. Others who made invaluable contributions included Major General H. P. W. Hutson and Malcom MacDonald (the British High Commissioner). Hutson's notes were published by the Delhi Birdwatching Society (established 1950) under the title *Birds about Delhi*, and Malcom MacDonald too published two works on garden birds.

The Delhi Birdwatching Society was established in 1950 by Horace Alexander and soon had an enthusiastic following. By the 1970s there were many ardent birders on its rolls, including Peter Jackson, Victor C. Martin, Captain I. Newnham, and the late Usha Ganguli. In 1970 the Society's species list for Delhi stood at 403.

The activities of the Delhi Birdwatching Society, however, seem to have tapered off some time after this, though it was briefly revived in the mid 1980s. At present, the recently established Delhi Bird Club is very active with 'bird walks' being organized every weekend to various places.

In the 1980s, the NGO Kalpavriskh conducted regular biannual bird counts in various areas of Delhi, and a monthly count at Sultanpur for eight years.

Today, there are probably more eyes following the flight of the shikra (and all of Delhi's birds) than ever before. The Internet has further encouraged the exchange of information, knowledge, and listings amongst birders and others, both in and outside the capital.[1]

Are birds in Delhi declining? The simple answer to this question must be yes, because we are rampantly destroying or encroaching upon their habitats. There was, for example, a wonderful wild and rocky landscape, studded with deep, water-filled quarry pits, lying between the colonies of Vasant Kunj and Vasant Vihar in south Delhi—an excellent bird habitat if ever there was one. The area has been earmarked for a multiple five-star hotel-building project and the birds will have to go. The Ridge, in north Delhi, has all but been converted into a landscaped park, attractively laid out perhaps to our city-bred eyes. But much of the natural undergrowth has been cleared away, making it inhospitable for peafowl and francolin to nest there and a not very interesting place for warblers and seedeaters like munias. Even migratory arrivals show declining trends and old-time Delhites complain plaintively that the wagtails, redstarts, and bluethroats that used to visit their lawns and gardens have all but disappeared. The number of winter waterfowl visiting the National Zoological Park too has drastically fallen in recent times; early in the 1980s, the main 'duck pond' used to be virtually 'house-full' with hundreds of happy, smug ducks. Today you can count them all too easily.

A more insidious enemy perhaps is pollution and pesticide poisoning. Birds love garbage, and landfill sites and garbage dumps are often great places to birdwatch. But what toxins mynas and starlings and kites, for example, ingest, while feeding off these places, we can only guess. The insectivorous species like the rollers and drongos seem to be declining as poisoning through pesticides takes its toll. Peafowl and *sarus* cranes also appear to have suffered the consequences of this unpleasant trend. Similarly water-birds

---

[1] Log on to Delhibird@yahoogroups.com, or check out the website of the Northern Indian Bird Network: www.delhibird.org.

and waders, feeding happily in the muck of the river must be ingesting at least some quantity of the chemical and faecal substances we heedlessly pour into the water.

Even so, Delhi still has much to offer the birdwatcher. It is up to us to see that the city remains a sanctuary for its astonishing bird population and species diversity. For a good place for the shikra to fly can only be a good place for us too.

Some Popular Birding Areas

As seen in the previous chapter, there are several areas in Delhi which are especially rich in bird-life and which can provide many rewarding hours of birding. Here a cross-section of them will be checked out in detail. The selection of areas has been done so as to cover most of the different habitats and ecosystems in Delhi.

The Jamuna at Okhla and the Surrounding Khader Countryside

Kalindi Kunj just north of Okhla Barrage on the west bank of the river makes an excellent starting point for a morning's birding at Okhla. You can get a panoramic view of the river and the backwaters caused by the Barrage, and their bird-life from the spurs of the high embankments jutting out into the river off the road along Fazal Enclave. In summer and during the monsoons, the river and its banks belong to resident species of waterfowl and waders. These include bitterns, herons, terns, cormorants (usually seen in flight), egrets, stilts, plovers (each of several species), and little grebes. For some years now, a flock of greater flamingos have been regularly seen on the mud spits and shallow water in the middle of the river.

In winter, the river plays host to a variety of waterfowl: northern pintail, shoveller, common teal, gadwall, wigeon, huge flocks of pochards (of at least three species), and coots. Migratory waders such as green sandpipers and redshanks scurry about the muddy edges, along with others of the salt-and-pepper long-legged wading brigade. A good viewing scope is essential if you are really to enjoy birding from here. The reed beds on the colony side of the road are alive with warblers, munias, babblers, and streaked weavers and are always worth checking out. On the downside, as this whole area is used as an open-air toilet, the stench on days when there is no breeze, or it blows the wrong way, can be quite nauseating. Also, don't look too hard at the bushes, first thing in the morning!

From Kalindi Kunj you can cross Okhla Barrage towards NOIDA and take the first left and drive northwards along the east bank of the river. A temple and cremation ground are located near the banks of the river a couple of kilometres down this road, guarded by a huge old banyan tree (an old, favourite meeting point for birders) usually full of screeching pied mynas and camouflaged green pigeons. There are fields (which often get flooded when the river backs up high behind the Barrage) on two sides, and the river, clotted with shiny water hyacinth and fringed with bull rushes and reeds, on the third. These lagoons and 'backwaters' that are formed by the Barrage can occasionally vanish altogether if the water is let out, leaving marsh and mud in their wake, which makes for a landscape full of surprises. In the fields adjoining the river, pied bushchats stake out their territories and natty long-tailed shrikes keep a sharp lookout from vantage points. Gangs of pied and bank mynas descend on the fields while exquisite blue-cheeked bee-eaters line up on the overhead wires like multicoloured notes of music. Amidst the reeds and hyacinth, the open-billed stork may stand stiffly to attention while chestnut bitterns go into hiding the moment they spot you. The tall reeds and bulrushes are aquiver with black-headed munias and warblers, and this is also where the black-throated weavers with their fog-lamp heads nest in great battalions.

Another rewarding option from Kalindi Kunj is to take the road going south off the NOIDA road (Road No. 13-A) just before

Okhla Barrage on the Delhi side of the river. Within minutes you will be in open agricultural country, surrounded by fields of aubergine and pumpkin (amongst other things). Dirt roads branching out eastwards lead to the river and its sandy banks, fringed with clumps of tall grass. This is excellent habitat for sand larks, small skylarks and crested larks, as well as francolin (both grey and black), quail, munias, warblers, and (during summer) nesting sites for small pratincoles. A road branching westwards towards Madanpur takes you through more farming country replete with drongos, bank mynas, shrikes, bee-eaters, bushchats, munias, warblers, and the like.

You can follow a similar route southwards from the west bank of the river into the khader, the flood plain of the river. This is good raptor country, with black-shouldered kites and kestrels (in winter) patrolling the fields, and you can even see the great tawny and steppe (this one in winter) eagles resting regally on mounds of mud.

Best of all, this area can make you quite forget that you are in Delhi at all.

The Northern Ridge, University Gardens, and Qudsia Gardens

The northern or Kamala Nehru Ridge in north Delhi, lies like a jagged emerald flint head between Civil Lines and Delhi University. This 87-ha, hilly woodland area has nearly entirely been converted into a landscaped park. Nevertheless, three distinct 'ecosystems' can be found here: the areas around the three water bodies or ponds; the areas that have been left 'wild'; and the areas that have been landscaped and planted with ornamentals. Nearly the whole of the northern Ridge is shaded by a decent thatch of acacia (of several species), neem, *peepul*, banyan, and other species.

The three most common species of birds seen here are the roseringed parakeet, the jungle babbler, and the collared or ring dove.

The areas around the water bodies are especially rich in birdlife. Bulbuls (redwhiskered and redvented), white eyes, kingfishers (that appear to breed here, though I have never discovered where), woodpeckers, barbets, coppersmiths, and, of course, roseringed parakeets are just a few of the species you can see in the thick foliage surrounding the ponds. Here too you may see the old little cormorant, or a pair of watchful pond herons. Whitebreasted waterhens stomp hurriedly under cover as you approach and every year raise broods of fuzzy black babies. I have spotted a pair of little green herons—rare for Delhi—standing on the banks of the filthy water body at the Hindu Rao Ridge. Unfortunately, the place is not very pleasant for birding as it seems to attract rather too many antisocial elements.

During winter, the pugnacious, grey-headed canary flycatcher may claim a pond for itself, driving away all competition, while dowdy chiffchaffs maintain a low profile amongst the crowded umbrella plants.

In the parkland tracts, magpie robins sing sweetly, hoopoes tend to the turf, and grey hornbills squeal in the peepuls as they stuff themselves with figs. Peafowl looking freshly enamelled, drift through the trees and then magically vanish, while babblers continue their clamorous hullabaloo amidst the leaf litter. In summer, the wild-eyed brain-fever bird (the common hawk-cuckoo) drives you crazy with its mocking 'Brain-fever! Brain-fever! Brain-fever!' call, and as the monsoons draw near, the koels bubble over with excitement as they hatch conspiracies to cuckold those street-smart crows.

In winter, the black redstart arrives, and lesser whitethroats 'Chttr-chttr' disapprovingly at you from the trees.

The wild, untamed areas, atangle with trip-wire creepers and spiky scrub, are the haunts of Indian robins and flocks of squeaky, white-throated munias. A shikra may suddenly dive out of cover, its eyes glittering hungrily. Many years ago, I saw a nightjar flitting amongst the hot rocks here, but only once. Grey francolin let out their ringing 'Pateela! Pateela! Pateela!' cries, but you really have to sneak up on them in order to spot them (and often they are up in the trees). Around thirty to forty species can be seen on a two-hour ramble on the Ridge.

Across the road from the burly Flagstaff Tower lie University Gardens—another very popular haunt for early morning walkers. In the big, gnarled trees studding the gardens you can be almost sure of seeing raucous Alexandrine parakeets and gravely suspicious, spotted owlets nesting in the holes and hollows. The feeding spots here are good places to see gangs of attractive, plum-headed parakeets, and any leafless tree is a sound bet for a flock of plump, yellow-legged green pigeons warming themselves in the early morning sun. Check out every dove in this area and you may be rewarded by the spotted dove which has been seen here.

The ancient and historic Qudsia Gardens near the Inter-State Bus Terminal have several grand old trees—neem, banyan, and peepul. Here families of alabaster-faced barn owls have nested for generations, and every spring you can see gentlemen grey hornbills cement up their mates inside the hollows of tree trunks and hefty boughs. Barbets and coppersmiths excavate neat, round holes in the dead wood of old trees and pied mynas build their scruffy, overstuffed nests in any available nook and cranny. In winter, the trees are festooned with hundreds of migratory black kites; silent visitors that add to the numbers of the residents kites that at this time of the year are squealing and mewling with passion and desire and constructing their colossal, twiggy edifices high up in the tree tops.

Buddha Jayanti Park, Lodi Gardens and Humayun's Tomb

Buddha Jayanti Park, carved out of the New Delhi Ridge is well known for the diversity of its trees. Its relatively open canopy makes it a good place for spotting birds as they fly across from one grove to another. Here you can get a clear, uncluttered look at golden orioles, small minivets, and brown-headed barbets. The areas along the borders of the New Delhi Ridge (a path runs alongside) can be particularly rewarding. It was here that I caught a glimpse of the rare (for Delhi) blue-headed rock thrush one morning in April, and followed a paradise flycatcher around as it fluttered to and fro amongst the garbage strewn all about. Also seen here was a tree creeper, with its squirrel's-back plumage and strange scuttling manner. During the monsoons, small, raffish parties of pied cuckoos can be seen winging across the Ridge, calling in their wanton, exultant manner. At the feeding spots, keep a lookout for rose-finches in winter that may join parties of munias, weavers, and sparrows pigging it out. During spring—in early April—the flame of the forest blooms, and attracts a host of birds.

Lodi Gardens may have become just a little too much of a celebrity garden for really enjoyable birding, but its trees and groves and monuments still provide shelter for a fine variety of birds. White eyes, sunbirds, wren-warblers flit around the flowering bushes, jungle and large grey babblers turn up the leaf litter (and discarded plastic bags) like ruthless customs inspectors, hoopoes tend the lawns, while spotted owlets peer out of their hollows in disapproval when visitors become too raucous, and parakeets streak across the skies screaming dementedly. One advantage of birding here is that the birds seem quite accustomed to lots of people and high decibel levels, and will allow fairly close approach. However, people- (and garbage-) loving birds like crows and mynas (and blue rock pigeons) have made these gardens a major *adda*, probably at the expense of other, more sensitive and aesthetic species, so you will see plenty of them.

The gardens in which Humayun's Tomb, and the tomb of Isa Khan are located are also worth a visit from the birder's point of view. House swifts flicker and squeak around the monuments, roseringed parakeets poke their bullet heads out of holes in the masonry, their colours matching those of the tiles, and peacocks dance with melodramatic flair from the tops of the gateways. Brown rock chats hop diffidently along the jagged stone walls, and a shikra may hide in ambush in a grove of frangipani. Bayas nest in the prickly date palms standing around the monument. The area is particularly rewarding during the monsoons. Next door, Sundar Nursery is another worthwhile area to explore.

The National Zoological Park

Delhi's extensive zoo has always been a place better for birding than for looking at its caged exhibits. The ponds and moats and acacia-clad island enclosures have long provided sanctuary to large numbers of birds of several species. During the monsoons, scores of egrets, cormorants, black-crowned night herons, and black-headed ibises nest in these acacias, and in August, the arrival of large numbers of painted storks livens things up further. They will nest here, and leave with their broods in the coming March. Other water-birds nesting here include white-breasted waterhens, Indian moorhens, and little grebes. A flock of wacky rosy pelicans have become a permanent feature (some of the birds were pinioned to draw the free-flying birds back) and may treat you to a take-off and short flight at around 10 a.m. when they set off on their rounds. Eurasian thick-knees (stone curlews) keep grim and silent vigil on the islands, in the company of more vocal red-wattled lapwings. Pied kingfishers may perform their spectacular spear dives into the water, after hovering from 10 m up, for what seems like several minutes as they perfect their aim.

In summer and during the monsoons, spot-billed and comb ducks (*naktas*) have the ponds all to themselves. In winter, migratory ducks like the northern pintails, shovellers, and common teal arrive, though their numbers have fallen in recent times (due to the stagnant, polluted water in the ponds, it is thought).

During summer, the groves of trees resound to the metallic calls of barbets and coppersmiths, and golden orioles and woodpeckers gleam enticingly as they play hide-and-seek in the canopy. In winter, red-breasted flycatchers and grey-headed canary flycatchers zip and flit amongst the branches; black redstarts eye you gravely, and white wagtails saunter on the grass like fat landlords. Migratory black kites too arrive in great numbers and wheel about the skies, silent and steely-eyed.

Plate 1 *White-rumped Vulture*   Disappearing from Delhi's skies; how much worse can omens get?

Plate 2 *Shikra*   Small pugnacious woodland hawk.

Plate 3 Greater Flamingos on the Jamuna doing the can-can.

Plate 4 *Spotted Owlet* Families are brought up in hollows and holes of large old trees; this one is a student at the north campus of the University of Delhi.

Plate 5 *Painted Stork and family* Colonizers of the zoo between August and March.

Plate 6 *Spotbill Duck* Commonest resident duck; not a romantic honeymooner!

Plate 7 *Magpie Robin*   Delhi's best songster.

Plate 8 *Ruff*   Migratory waders that pass through Delhi in large smoke-like swathes.

The Sultanpur National Park and Badkhal Lake

Sultanpur National Park, 15 km from Gurgaon, was during the 1980s and early 1990s, one of the most remarkable shallow wetland bird habitats in the country. Just 1.5 sq km in extent, the area attracted nearly 250 species of birds, of which approximately 110 were winter migrants. The *jheel* (lake), which had been attracting birds for over a hundred years, lies in a shallow natural depression in which rainwater collected. In winter, the shallow waters and muddy banks attracted thousands of migratory waterfowl—ducks of around thirteen species and both bar-headed as well as greylag geese. Large flocks of common and demoiselle cranes used the place as a nighttime shelter (raiding the surrounding fields during the day) and thousands of ruff and reeve stopped over for a bit of rest and recuperation en route to their final destinations further south or back home north. Jaunty flotillas of rosy pelicans would fish the waters in the years when the monsoons had been good. Raptors like the marsh harrier, long-legged buzzard, lesser spotted eagle, kestrel—and once even an osprey—had a good time of it as they coursed the lake and surrounding grasslands, though the resident black-shouldered kite was none too pleased with the non-resident interlopers!

Amongst the resident species, the graceful sarus crane nested here during the monsoons, and the increasingly rare black-necked stork would bring its gangly progeny here for an outing. Lapwings and black-winged stilts as well as egrets of all four kinds, proliferated.

The acacia woodlands surrounding the lake provided nesting sites to large numbers of red turtle doves who would arrive just for this, in peak summer. These thorny acacias also provided refuge for babblers, warblers, flycatchers, mynas, parakeets, koels, magpie robins, and a host of other woodland birds. On the grassy tracts, munias and warblers swung on the reed stems and the shrikes kept vigil while hard-eyed, satin-clad egrets stalked the ground relentlessly. Bee-eaters and swallows scissored the skies, while orioles, doves, and drongos nested in the trees in the grounds of the tourist complex.

But then the lake dried up due to a combination of natural and man-made factors. Large new farms in the surrounding area sucked up a lot of the groundwater and prevented the natural inflow of rainwater into the basin. The water table dropped alarmingly and then the rains failed. For five disastrous years from the mid-1990s onwards, Sultanpur lay bone dry. Migratory birds stopped coming and doomsday seemed nigh. A plan has now been implemented to get water from a nearby canal. The idea is to regain as closely as possible, the water levels that existed in the lake during its halcyon years. Whether the plan is properly implemented and succeeds, and the birds return, remains to be seen. Even so, Sultanpur would be interesting to visit, if only to check on how its grand rehabilitation plan is working out.

Badkhal Lake, adjacent to Faridabad, is a deep-water body located in the midst of rough Ridge country. As a popular tourist resort, it gets noisy and crowded during the weekends, and so is best visited very early in the morning. In winter, you can see the 'deep-water' ducks here: common pochard, tufted pochard, and the flamboyant red-crested pochard. The fish tanks of the Fish Farmers Development Agency on the inland side of the bund are good places to look out for blackwinged stilts, lapwings, kingfishers, cormorants, sandpipers, and redshanks that take advantage of the inventories!

The rocky landscape around the lake is studded with stunted (because they are lopped) *dhak* (flame of the forest) trees, which during March and April burst into bloom. This attracts scores of mynas, parakeets, white-eyes, sunbirds, and green pigeons. The more open, rock-strewn areas are the haunts of yellow-throated sparrows, wood shrikes, and Indian robins.

Open Skies, Open Seasons

A question that I am often asked is, 'When is it a good time to see birds?' or 'Is now a good time to see birds?' To these my answer has always been, 'It's always a good time to see birds,' since there is something or the other going on in the avian kingdom right through the year!

Delhi has fairly well-demarcated seasons. Winter creeps slowly into the capital from November onwards (especially after Diwali), bites hard by December–January, and melts away by the end of February. A brief spring emerges, lasting through March (Holi is usually the turning point), before summer, with its hot halitosis *loo*s, comes roaring in like an unleashed dragon. The heat and humidity reach cauldron levels by the end of June and beginning of July, before the monsoons break and bring blessed relief. By September the skies are clear again, and though the sun is strong it lacks the blowtorch ferocity of summer. There is dew on the ground at dawn and, after Diwali, the ground mists creep over as winter draws near.

As far as birds (and birders) are concerned, there are really two main seasons: the non-breeding season for resident birds and migratory season for non-residents, and the breeding season for resident birds. The migratory season begins in August–September and lasts till March–April, and the breeding season takes off in February–March and lasts till September–October. Let us start our journey through the year at the time when this is being written—

early May. Summer has just begun to stoke up the heat, and several resident species have got down to the grim business of nesting. Bank mynas have taken apartments in the weep holes of walls and are busy furnishing these with feathers and rags. Coppersmiths, brown-headed barbets, and woodpeckers will have finished chiselling neat, round holes in soft deadwood, and will have settled in. Actually, holes and hollows in tree trunks and branches are at a great premium at this time of the year as a whole host of species—mynas, owls, parakeets, magpie robins, barbets, woodpeckers—find them excellent nesting sites. Often you can witness fierce property disputes in progress in any Delhi park, between both members of different species and those of the same species. The niches and crannies and cracks in many of Delhi's historical monuments too provide nesting sites for several species like swifts and rock chats.

Green bee-eaters and kingfishers too will have started nesting by now. They excavate tunnels in earth banks and so must raise their broods before the monsoons break and turn their homes into mud. So must birds like the paddyfield pipit that may construct its 'nest' inside a dried hoofprint in the mud. That smart upright standing plover, the red-wattled lapwing, will have been having hysterics for the past month or so by now. Lapwings lay their eggs on the ground and go ballistic if anyone approaches even remotely close. The smart black-winged stilts that potter about on the river bank or along the edges of any and every *ganda* nallah or ditch, will also have commenced nesting, and are equally hysterical, launching deadly dive-bombing attacks if you get too close to their nests. In the water, the resident spot-billed ducks will be courting—the mutual bobbing of heads is followed by some rather rough-looking honeymooning. The sandbanks and reed beds along the river are also nesting sites for terns, plovers, and pratincoles.

Summer, too, is the time when crows get high blood pressure. Delhi's green canopy will be filled with the blood-eyed, sin-black koels with their dappled brown wives by now, chasing each other madly through the trees, calling with maniacal glee. The cryptic-coloured brain-fever bird, or common hawk-cuckoo, that has been calling *ad nauseum* since March, will also have started diddling the scowling jungle babblers, like sophisticated city slicks taking simpleton cops for a ride. The mellow flutings of the golden oriole are, by contrast, a lovely call to listen for at this time. These beautiful gold and black birds arrive in tree-shrouded parks and gardens by April and commence with courting and nesting. Neem trees are a great favourite with them, especially if there is a black drongo in residence in the same tree to provide them with protection.

Late April and early May are also when screeching mobs of streaked weavers begin exploring nesting sites on the Ridge and the reed beds adjoining the river. Bayas too begin constructing their beautiful, vase-like nests, which in due course will be examined thoroughly by visiting females. Spring and summer are also the seasons for fruit and flowers, and many species are dependent on these for food.

As the monsoons approach, the nesting season gains momentum. The explosion of insect life during the rains ensures a sufficient supply of high-protein fare for thousands of ravenous baby birds. Warblers, tailor birds, babblers, bulbuls, munias, white eyes, bushchats, and bayas get busy. Already, the herons and egrets have decked themselves out in their best, and have begun courting and nest building. Egrets, in fountains of lace and with the eyes of gin-drinkers, shimmer and tremble as they greet each other at their nests. You can watch and monitor their progress at Delhi zoo, right from the courting to the first flight of the brood. In the marshy tracts along the river, pheasant-tailed jacanas trip lightly over the floating vegetation, to and from their nests, and the dark, brooding ponds on the Ridge play host to stentorian-voiced white-breasted waterhens that soon are taking black, fuzzball babies for walks over the water lilies. Resplendent in wine and burgundy, little grebes potter about at the edges of ponds and water bodies, letting off their trilling alarm clock calls from time to time and then diving beneath out of sight.

The imminent arrival of the monsoons is heralded by that wild, exultant creature, the pied cuckoo. Rumour has it that this bird flies across all the way from East Africa for the express purpose of courting wantonly at Buddha Jayanti Park or the Ridge, and disposes its responsibilities in the nest of some poor duffer babbler. The monsoons are a time of plenty but a time of danger too. Whooping ghoulishly, the russet and coal black greater coucal searches through the lush foliage for baby birds, like some monstrous bogeybird doing the rounds. Crows and tree pies too are not averse to searching out and destroying nests. They too have families to feed.

By August, if all goes well, the painted storks will have arrived at Delhi zoo (and been ecstatically greeted by the press as 'Siberian cranes'!) and will have started nest building, alongside the egrets, cormorants, ibises, and night herons already in residence. They will remain here perhaps till the following March, by which time their families will have grown and be ready to leave home.

Now too, in some of Delhi's parks, you may be startled by a sharp irritable 'Che-chwe!' call. Look carefully and you may spot the elusive paradise flycatcher. Actually this exquisite bird may be seen as early as March and April, though its movements and breeding habits remain a bit of a mystery. By August or September, the male will have shed its magnificent tail streamers, but none of its restless charisma.

As the monsoons retreat the first of Delhi's non-resident avian guests begin to arrive. By the end of August, or even earlier, the first of the ducks will have begun to arrow overhead and freckle the surface of the river or large, open water bodies. The gargeny— or blue-winged teal—is amongst the earliest to land, followed swiftly by the common teal, pintails, and shovellers. The deep-diving pochards and geese arrive a little later, by November and December, as winter begins to bite. By now vast numbers of long-legged waders—redshanks, sandpipers, lapwings, ruff and reeve, snipe, and godwit—will also have arrived. Some, like the ruff and reeve, leave for destinations further south. November is probably the best month for viewing the maximum diversity of migratory species. Nonetheless, all through winter, the river plays host to thousands and thousands of waterfowl and waders. These birds will have been followed down by their nemesis—eagles, buzzards, falcons, and harriers—that give the massed flocks the frights by quartering low and slow over them, as though carefully surveying items on a menu (which is exactly what they are doing!).

Another regular feature is the arrival of large flocks of brown- and black-headed gulls that blizzard around the pontoon bridges and *ghat*s on the river. They are assiduously fed every day and come winging over swiftly when called.

The small grassland and woodland birds will have arrived too. Grave black redstarts, bluethroats, and wagtails grace parks and gardens and the ubiquitous acacia canopy quivers with irritable lesser whitethroats, forever 'Tch-tching!' their disapproval. Red-throated flycatchers toss themselves after insects in parks and gardens, and in areas adjoining water bodies the pugnacious grey-headed canary flycatcher guards against all comers. Out in the more open and scrubland tracts, collared bushchats keep watch and dun and biscuit warblers of various kinds flit restively from bush to bush.

By February there will be a restiveness in the air. Feeding frenzies will have commenced and the ducks will have begun making practice sorties. There will be a sudden influx of birds like, godwits and ruff and reeve, passage migrants from the south. All through March and April the exodus will continue. Ducks on the river one day, will suddenly be gone the next. Swallows will congregate in great chattering flocks on transmission lines and, by the end of April, peepul trees in many areas, (even the railway station) will be teeming with aggressive rosy starlings pigging it out before embarking on their return journey.

By now, however, the resident magpie robins begin their flute concertos and the shimmering purple sunbrids are zipping and shrilling everywhere like pop stars on a high. In the big, ancient trees of Qudsia gardens the antediluvian grey hornbills start their quaint courting, the male offering his beloved, neem or other berries as though they were emeralds. Besotted brahminy starlings throw back their heads and sing with all the passion they can muster. Red-vented and Red-whiskered bulbuls spread cheer all around with their musical, mood-elevating whistles. Out in open scrub country, the larks start performing; small skylarks launch themselves off tussocks, climb steeply and fluttering in tight wavy circles, and sing non-stop for longer than you can keep looking up at them without blinking. Even the harsh-voiced shrikes produce surprisingly mellow melodies at this time. The peafowl start dancing early: I have seen young males with half-grown trains, strut with open fans in front of peahens in January. The hens, in time-honoured fashion, ignored the puppies. But later, when full-grown and magnificent, I have watched the males dance all through the day in July and also at locations as dramatic as Humayun's tomb. In the fields, the sarus cranes will also be smitten, and take to dancing, prancing and braying in the wonderfully unabashed manner of their species.

Alas, Delhi's spring is all too short. And all too soon, the time for singing and dancing and courting is over. Suddenly, in the nests, there are mouths to feed and stomachs to fill.

Field Notes from Three Typical Birding Trips

˯

The Ridge, 6 November 2000, 7.30–9.45 a.m.

A little too hazy for good photography, but at least the sun was out. Smoky cool. Birdcalls at the entrance (Rajpur Road gate): plumheaded parakeets streaking somewhere overhead; brown-headed barbets trying to decide whether or not to call; jungle babblers having a fracas in the undergrowth; white-throated kingfishers, giving tongue. A number of lesser white-throats calling from the canopy.

At Khooni Khan jheel, heard the red-throated flycatcher—saw it flit high in the peepuls. Then, at the Serpentine, heard the grey-headed canary flycatcher in song, which always reminds one of the hills. Of course the little fellow didn't show himself.

Area around the Serpentine is full of zipping and squeaking sunbirds, kept company by white-eyes aplenty. The 'wild' flank didn't have very much on offer—lesser white-throats and jungle babblers mostly, but then I hastened through it. Burst of bulbul song at the nursery end. Then, in the area between the wild flank and shadowy parkland (the monkeys' 'Bada Nashta Khana'), a soft 'Wheep wheep!' call. Sparrow-sized bird, ashy grey on top, with canary yellow underparts and patch on wing. It was flitting after insects and uttering these sad sounding 'Wheeps'.

Seems to be a single female long-tailed minivet—just where are the males?

Dull green warblers at the silk-cotton-grove end of the BNK, near the Serpentine. Also a female redstart and the usual complement of lesser white-throats and white-eyes. Spent a while photographing white-breasted waterhens at the feeding spot on the path beside the Serpentine. Seems to be a family of three—they're not very troubled by the walkers and joggers. Would scurry off and return after the disturbance had passed.

Then, in flew a flamboyant white-throated kingfisher, posing wonderfully on the dead tree, the sun on its flanks. A squirrel tried to dislodge it, even by approaching from under the branch and nipping its tail, but with little success. But then a monkey gang showed up with crows in tow and that was that.

The following were observed:

1. Shikra (heard, 1)
2. Brown-headed barbet (heard, 2–3)
3. White-throated kingfisher (1)
4. Plum-headed parakeet (heard, 1)
5. Jungle babbler (10+)
6. Red-vented bulbul
7. Red-whiskered bulbul (flock of 12 or so)
8. Ashy prinia (1)
9. Tailorbird (1)
10. Magpie robin (1, male)
11. Rose-ringed parakeet (ubiquitous!)
12. White-breasted waterhen (family of 3, including bratty junior)
13. Lesser white-throat (at least 10)
14. Dull green warbler (1)
15. Long-tailed (?) minivet (1, female)
16. Laughing dove (5+)
17. Eurasian collared dove (5+)
18. Black kite (about 5)
19. House crow (<10 party)
20. Red-throated flycatcher (<5, one seen, rest heard)
21. Grey-headed canary flycatcher (heard, at least 2)
22. Oriental white-eye (bunches! plenty!)
23. Black redstart (1, female)

Field Notes from Three Typical Birding Trips ◆ 31

Zoo, 2 December 2000, 9.40 a.m.–12 noon

Cool, clear, and crisp—too good to miss. The zoo has just a smattering of ducks, mainly pintails. A few shovellers, and saw just three common teal. Dismal really—in comparison to old 'house-full' days the main pond was nearly empty today.

Painted storks are nesting in strength—lots of morose, unwashed-looking youngsters standing around. There appear to be more nests on the islands around the pelicans' pond. Eight or nine moorhen swimming near the island of the main pond, as also similar number of red-wattled lapwings holding a silent (prayer?) meeting next to the blackbuck's enclosure. Spotted four or five thick-knees (stone plovers) playing statues on the island in the main pond.

Pelicans were grooming themselves thoroughly, scattered in three or four groups all around the pond. Some remain pinioned. Only one took off and flew as I watched, the rest stayed back bobbing about like boats, and getting their feathers into proper trim.

The giraffes were necking languidly as a crowd of schoolkids screamed and shouted excitedly. They licked each other with their blue grey tongues. The shorter (younger) of the two seemed a little restive; its partner stood stock still, looking infinitely sad—and dreaming about National Geographic films on Africa?

The pond opposite the chimps' enclosure had some mongrel domestic ducks and geese and a few pintails. Interesting goings-on at the far end (towards the hippo enclosure): Dried fish had been laid out, and helping themselves to it were adjutant storks, egrets, night herons, black-headed ibises, pond herons, and three cats! The cats would crouch at the edges of the feeding area, slink up, and snatch a mouthful, ever-mindful of those stabbing bills and spiky feet. Every time they came forward, the birds would instinctively spring up and backwards, before stabbing forth again and making the cats back off. At one stage, however, the cats and egrets were eating together in a state of uneasy truce, from opposite ends of the 'table'. When scared off by workmen, the egrets were off in a blizzard of wings, and now the pond herons (who were probably not getting their fair share) nipped in deftly and went gobble-gobble-gobble. The cats too were quick to recover and stuff their faces. Then of course the egrets came drifting back and the delicate balance of power was restored. Seems like the cats preferred fish to fowl, though, admittedly, tackling a well-armed egret or heron is obviously far more risky than helping yourself to fish served up on a platter.

Okhla, 18 June 2000, 6.00–8.30 a.m.

Day of the bitterns! Sultry but clear, if a little vapoury with a faint breeze from time to time. Met on the west bank down the road from Kalindi Kunj.

First to be sighted, a yellow bittern, flying, (a little more hastily than the pond heron), black wings and tips, flat, blacktopped head, ochre-brown in general appearance. In the course of the morning saw plenty more of them, all in flight and in pairs. Focused on one that had landed on hyacinth through the binoculars, clear and sharp with yellow-ringed eye staring in demented heron manner. Also in the hyacinth, a pair of ever-elegant pheasant-tailed jacanas looking quite spiffing. Streaked weavers busy nesting in the reeds everywhere. Also pretty ubiquitous, pied bushchats with families in tow. Baby bushchats are darkly marked and streaked and have that wide-eyed, innocent expression and bug eyes.

Drove across to the eastern bank and down the narrow black road along the river. Stopped awhile for a spot of tern watching. Apart from the river and whiskered tern, saw the little tern too; it has a more fluttery flight than the others as though it has to work harder to stay airborne. Finally arrived at the famous banyan tree. Fields surround the area on two sides and the river on the third. Spit of land extends into the river. Hyacinth clogging up towards the banks. Spotted a very clean and white looking open-bill stork here. A pair of yellow bitterns dived into the vegetation and vanished. Other birds spotted included night herons and pond herons.

Streaked weaver replaced by black-throated weavers, nesting in the bull rushes. Also, streaked babbler, along with the common babbler. Streaked swallows lined up on the wire as though for inspection and one lovely black-headed munia (in paradise flycatcher chestnut) showed itself briefly before vanishing.

Wandered beyond the cremation ground, preceded by a pair of spotbills who eventually took to the water. Then, a pair of lovely chestnut bitterns. They are a rich rufous saffrony colour, absolutely unmistakable. They were also in flight but too far away for photography. A flight of black-winged stilts flew overhead, also spotted were some avocets.

Then flamingos. A flock of over one hundred took to the air for a flypast. Long, slim, ballerina like, the group broke up, one-half flew downriver—southwards—the other half northwards. And one silly fellow honked dolefully in the middle, not knowing where to go, before joining the northbound group. They have a very goose-like honk.

Also in the distance, a handful of lesser whistling ducks, but looking better fed now than in winter. They were also flying less swiftly than I remember them doing (because they've put on weight?)!

Purple herons also present, flying slowly past, or standing stock still in the reed beds. Also a large flock of spotbills. Caught just a shadowy glimpse of the black bittern here, as it flew—that made it three bitterns for the day!

Crow pheasants (coucals) inspected the reed beds and one Indian moorhen took a long circuitous flight over the reeds. Lapwings, of course, went berserk over the dogs.

The banyan was host to pied mynas and green pigeons and on a tree nearby a long-tailed shrike gave breakfast to two babies, who were old enough to have left the nest but still demanded to be spoon-fed. Then, on the wires, blue-cheeked bee-eaters looking more beautiful than I had ever seen before. In the pale, silvery light, their heads were a satiny turquoise, set off by the dashing dark mask. The throat was canary yellow with a splotch of dark saffron on the breast, almost carmine, followed by the matt grass green plumage. The poor little green bee-eaters were outclassed completely.

Even the ashy prinias were looking great at this time, and shouting loudly. But the reeds really belonged to the weavers and bushchats. Also present on the water, dabchicks and, of course, cormorants and one kingfisher.

II
THE ROGUES' GALLERY

The Birds

Grey Francolin (Grey Partridge), *Francolinus pondicerianus*, Teetar

The cheerful ringing 'Pateela! Pateela! Pateela!' calls of the grey francolin are a familiar early morning sound in many of Delhi's more rambling parks and gardens. About the size and shape of a half-grown chicken, this dumpy, stub-tailed, scrub-loving game bird appears more brown than grey, with chestnut cheek patches and throat, a whitish eyebrow or supercillium, and delicate black and buff vermiculations.

However, grey francolin are cautious and wary birds, difficult to spot and quick to disappear. More often than not they will suddenly break cover from virtually underfoot, with a startling 'Bhrr!' and be gone before you regain your composure, let alone focus on them. They also like to roost up in trees and can, therefore, also give you a fright by blasting off an overhead branch and careening crazily away, to vanish uncannily into the undergrowth. (They are strong runners too). The challenging 'pateela! pateela! pateela!' calls are made by the males (who wear spurs on their legs) and may be answered by the hens' gentler sounding 'kella! kella! kella!' replies. A throaty 'Khirr! khirr! Khirr!' is a warning call indicating a possible, imminent blast-off from cover! However, if you are quiet and sneaky, you can sometimes come across parties of two to eight birds at feeding spots in parks—like

Buddha Jayanti Park or the Ridge—after the early morning crowds have left. The New Delhi (or central) Ridge, the northern Ridge, the badlands around Tughlakabad, the campus of Jawaharlal Nehru University, the rocky Ridge areas of south Delhi, and the fields at Okhla are other favourite haunts.

The birds also set forth on their perambulations in the evenings, scratching the ground for grain, insects, and seeds. They are believed to nest throughout the year, scratching out shallow, stone-lined scrapes in the undergrowth. Four to eight milky-coffee-coloured eggs are laid.

The birds have been extensively trapped and hunted—a practice that still illegally goes on occasionally, on the New Delhi Ridge for example. They are also kept as fighting birds, and bouts are often organized in the Walled City. With the extensive clearing of the undergrowth in places like the northern Ridge, they are finding suitable nesting areas harder to come by. Delhi would indeed be a grimmer, duller place without these sprightly stalkers of the undergrowth.

Black Francolin (Black Partridge), *Francolinus francolinus*, Kala Teetar

This one is an enamelled beauty: the same size as the grey francolin but finished in glistening black on top, polka dotted and streaked in white, with the feathers rimmed in gold. The abdomen and flanks are chestnut. A copper band around the throat and a gleaming white cheek patch complete its stunning ensemble. The hen is a paler version of the cock, without the collar or ear patch.

Much less common than the grey francolin, the black francolin can usually be heard uttering its creaky 'Chik-cheek-cheek-keeryak!' call early in the mornings or late in the evenings in well-cultivated well-watered country, like in the fields south of Okhla Barrage or those around Sultanpur National Park. You will be very lucky to actually spot these elusive birds, though sometimes they do perch on—and call from—the bare branches of tees. Their calls have a strangely ventriloquist quality that can make it difficult to pinpoint the location of the birds even when they call from nearby.

Black francolin nest between March and October in fields, dense scrub, and bushes. Six to eight pale olive to chocolate-brown eggs are laid in a shallow, grass-lined scrape. These birds too were extensively hunted and trapped in and around Delhi. If you are missing out on seeing these birds, you can take the easy way out and admire them at the National Zoological Park.

Indian Peafowl (Common Peafowl), *Pavo cristatus*, Mor

With their dark eloquent eyes, shimmering, palpitating throats, and bejewelled tiaras and trains, peafowl mince gracefully in parks and gardens and around monuments everywhere in the capital. Common enough not to warrant much attention, they can still evoke moods of romance and nostalgia in the right setting and weather. Let a storm rumble up in torrid mid-May, with a blustering wind and great drum-rolls of thunder and the peacocks will exult, screaming 'May-yew, may-yew, may-yew!' as they scramble up to the tops of the lemon-gold laburnums, waiting ecstatically for the first big drops to splash down. Watch a peacock dance on the gateway of Humayun's Tomb, against a backdrop of glowering monsoon granite, or two magnificently tapestried males duel for a harem in the dark cloud-light of Nicholson Cemetery and you will be transported back in time, two hundred years in an instant.

The Ridge, Deer Park, Buddha Jayanti Park, and probably every large rambling park, green area and rock-strewn wilderness in Delhi supports its fair share of peafowl. Usually unharmed, except by those who hunt them for their plumage, peafowl have even taken to nesting in residential areas: on terraces, porticos, and pelmets of houses and buildings in many of Delhi's greener residential colonies. They breed between January and September, and some of the best and longest dance performances may be seen in July and August. I have watched a cock dance tirelessly through the day in Nicholson Cemetery while his harem of six hens tried hard not to appear impressed. The hens too have a hierarchical peck order and probably it is the senior most hen that gets to mate with the male. The males are so competitive they can't even stand the sight of their own reflections in the hubcaps and windscreens of cars and will peck furiously at them if so confronted. One peacock was so enamoured of his position on the bonnet of my car that he only got off when I got in and started the engine.

Peafowl roost high up in the trees, and plane down (honking madly) in the morning, to begin their search for shoots, insects, grain, lizards, and even snakes. They may also enter private gardens on the sly and make short work of lovingly planted seedlings and young shoots. Complete silence is maintained during such raids. In the surrounding countryside, crops are similarly targeted.

Greylag Goose, *Anser anser*, Rajhans

This large (75–90 cm), cinnamon and grey goose has distinctive flesh-pink legs and bill, and is thought to be the ancestor of all the domestic breeds of geese. A winter migrant to large water-bodies in north India, the birds can be spotted on the mudbanks of the Jamuna and in water-bodies around the capital like Sultanpur. Their far-reaching 'Aahng-ung-ung!' calls are one of the most evocative sounds floating down from a blue winter sky, as they fly in great wavy echelons or classic 'V' formations to raid fields of winter crops. The birds are exceedingly wary. They arrive by around November and are gone by the following March.

Barheaded Goose, *Anser indicus*, Rajhans

Perhaps more commonly seen around Delhi than the greylag, the barheaded goose is a handsome bird in silver-grey livery, with a white head streaked with two charcoal bands, a black-tipped, yellow bill and papaya-orange legs and feet. Slightly smaller than the greylag, at between 70 and 76 cm, it is also a migrant and may be seen on the sandbanks of the Jamuna. Large flocks of these birds could (in the good old days) also be seen at Sultanpur; grazing in the grassy areas, snoozing in the soft mud, or bobbing on the water, head-in-wing on a buttery winter morning.

Flocks can also be flushed from the surrounding wheat and mustard fields, rising en masse with a great honking clamour, and making off at good speed. The arrival of a squadron over the lake is always a treat to watch as they curve and bank low in formation over the water, black and white crosses against the dark green background, before straightening out and splash landing with contented gabbling. Sometimes, when flying high, they jink and sideslip and drop alarmingly, appearing out of control, but then regain their composure to make yet another perfect landing. Like the greylag, their musical 'Aang-aang!' calls stir up strange, primordial feelings and are quite unforgettable. Barheaded geese are also extremely wary. They breed in the cold, high-altitude lakes of Ladakh and Tibet between April and June. Both these species of geese are thought to overfly the Himalayas while on their migratory journey to and from the subcontinent.

Lesser Whistling Duck (Lesser Whistling Teal), *Dendrocygna javanica*, Seelhi/Seelkahi

A small (42 cm), swift, roast-brown and maroony-chestnut resident duck, with fast-fluttering flight and a spirit-lifting 'seasick, seasick, seasick,' or 'Whi-whee! Whi-whee! Whi-whee!' whistle. It is considered to be rather uncommon in the Delhi area, with few records going back for the last one hundred years. However, the birds have been seen in recent times, both on the Jamuna at Okhla, as well as at Sultanpur.

Lesser whistling ducks like spending the day resting up in weed-covered ponds or jheels and foraging at night in nearby paddy and crop fields. They are not averse to perching on trees, and often nest in the hollows of tree trunks.

A few years ago, I got a good close look at one in the duck pond of Delhi zoo and was able to properly appreciate its glazed roast plumage (and big, black, flippered feet) and sprightly, upright stance and manner. I couldn't tell whether the bird was pinioned or not, but I never saw it on subsequent visits so can only hope it was free and had flown away.

Ruddy Shelduck (Brahminy Duck), *Tadorna ferruginea*, Chakwa/Chakwi/Surkhab/Lal

A large (61–7 cm), saffron-copper, goose-like duck, with a pale caramel head and striking bottle-green, black, and white wings. The male wears a black collar, which in winter (when the birds visit) may be faded and indistinct. The female has no collar and her head is much more blonde.

Small to large parties of ruddy shelduck, sometimes numbering about one hundred birds, may be seen grazing or lazing on the mudbanks of the Jamuna right along its course through Delhi. They make a lovely sight while in flight but are usually wary, solitary birds that seem to prefer their own company. They arrive around October and stay on till April.

Delhi zoo has several pinioned birds that behave somewhat like evangelistic guardians of morality as far as the other ducks (and water-birds) are concerned. Having done this duty they might also sidle up to you for a tidbit, honking self-righteously. Ruddy shelduck nest in Ladakh, Tibet, and Nepal between April and June.

Comb Duck (Nakta), *Sarkidiornis melanotos*, Nakta

This large (56–76 cm), inky, blue-black and white duck is more often seen in flight than on the ground (or up in a tree). The head is white, spattered with inky spots, and the male, who is much larger than the female, carries a black 'comb'—a fleshy knob—on its black beak. The wings are black, glossed with blue-green; the fuselage is white.

I have seen groups of naktas congregating on the islands in the ponds of Delhi zoo, looking quite stuffy and officious. Small flocks comprising nine or ten birds can also be seen flying around Sultanpur. Naktas are resident ducks that breed during the monsoons and nest in the hollows of trees usually close to, or standing in, water.

Cotton Pygmy Goose (Cotton Teal), *Nettapus coromandelianus*, Girri/Girria/Girja

The smallest of our ducks (31–7 cm) and a resident, but not much is known about its status around Delhi. It has been seen at waterbodies around the capital as well as at Sultanpur, in summer and the monsoons.

The male has a white face and breast, and wears a black skullcap and breast band. His upper plumage is greeny-black, the flanks and underparts a smoky blue-grey. The green-glossed wings have a broad, white bar on their trailing edges. The female, overall, is less striking and more peppery-grey and lacks the black cravat of the male.

I have seen this rather quiet and downcast-looking duck at Sultanpur during the monsoons, but it appears to be one of those introverted little birds who would rather not be stared at, and floats away under cover when it senses it is under observation. It breeds during the monsoons, nesting in the hollows of trees standing near water.

Gadwall, *Anas strepera*, Myla/Beykhur/Bhuar

One of the commonest of our migratory ducks as well as one of the most soberly dressed. The drake is clad in conservative blanket grey, finely patterned in black and white and has a velvety-black rear end, dark-grey bill and whitish belly. A chestnut patch in front of a black and white speculum is visible while the bird is at rest. The duck is a mottled dark and light brown.

Large flocks of gadwall can be seen alongside other ducks in the Jamuna at Okhla in winter. They usually arrive by the end of October.

Eurasian Wigeon (Wigeon), *Anas penelope*, Peasan/Patari

An attractive silver-grey duck (45–51 cm) with a finely pencilled plumage, salmon-pink breast, auburn head crowned with pale creamy-gold, and a slate-grey bill. The female has a brown head (and no golden crown) and a prominent white belly and russet flanks but, frankly, is quite ordinary looking.

Wigeons are migratory and may be seen on the Jamuna and water-bodies around Delhi, though I have not as yet spotted them at Delhi zoo. Most arrive by early October and are gone by March. The drake is said to utter a pleasant 'Whee-yoo!' whistle.

Mallard, *Anas platyrhynchos*, Nilsir/Nir Rugi

About the size of a domestic duck (50–65 cm)—which have mostly descended from them—the mallard has a big emerald head and bright-yellow bill and two delightful black kiss-curl feathers sticking out of its tail. A narrow white band separates the head from the plump chestnut breast. The rest of the plumage is greyish, finely woven with black, the feet are bright plastic orange. The speculum is a lovely purple-blue and especially attractive in flight. The female is mottled, in buffs and browns.

I have seen mallards fairly frequently at Sultanpur in winter, though in the past—during the 1960s for example—they were not recorded very often from the Delhi region. With their glistening green heads and purple-blue speculums flashing in the sun, they make an attractive sight as they wing past. You can, of course, appreciate their plumage at close quarters at Delhi zoo where they are kept (and maybe hybridized) as exhibits.

Spotbilled Duck (Spotbill Duck), *Anas poecilorhyncha*, Garm-pai/Gurgal

A plump, busty, salt-and-pepper, scaly-patterned duck (58–63 cm) with a bright-orange *bindi* at the base of the forehead (and matching legs) and yellow tip to dark bill. The speculum is white and metallic green—or purple, depending on the sub-species. The top of the head is dark brown as is the region around the eyes, and which makes the bird look as though it has applied kohl and gives it a somewhat soulful appearance.

Delhi zoo is probably the best place to get a close look at these resident ducks; the birds are free flying and breed here. I have watched them court and mate in the ponds—not a particularly tender spectacle as the male appears quite content to drown and rape his partner all at the same time. Fairly large flocks can be seen on the Jamuna at Okhla as well as at Sultanpur and Badkhal, and probably every self-respecting water body around Delhi. The birds breed during summer and the monsoons. Home is a wad of grass and weeds set down in squelchy places near water, and a dozen eggs may be laid. The sexes are alike.

Common Teal, *Anas crecca*, Chhoti Murghabi/Kerra/Lohiya Kerra

This pint-sized little duck (34–8 cm) is one of our most common migrants. It has a finely woven greyish plumage and a distinctive caramel patch on its flanks. The speculum is black, green, and buff, visible in flight—which is rocket swift. The rich russet head is dominated by a broad green mask that runs across the eyes to the nape and makes this little duck look as though it is all set for a costume party. The green mystery mask can also turn royal purple depending on the angle of the light. The female is mottled in pale and chocolate brown and has a green speculum.

Though common teal are pretty abundant on the Jamuna, at Sultanpur, and other water-bodies, it is usually difficult to get a close look at them as they tend to be shy and wary. Often they lie low in the tawny grass or muddy banks, and will suddenly go careening off like mad missiles when you get too close. However, you can get a good look at them at Delhi zoo where they keep visiting shovellers and pintails company. But even here I have found them to be suspicious and shy, and they will sidle away the moment they realize you are taking a little too much interest in them for comfort. You have to sneak up to them unobtrusively and then pretend that you haven't noticed them at all. Usually they arrive by September and leave by the following April.

Gargeny (Blue-winged Teal), *Anas querquedula*, Chaita/Khira/Patari

Somewhat larger than the common teal, the gargeny (37–41 cm) has a pinkish-brown speckled head and a prominent white eyebrow or supercillium, which gives it a bit of a grandfatherly appearance. The breast is greyish and the flanks are brown. The forewing is a beautiful flame blue, the speculum green, with its trailing edge white.

Gargenys are early-bird passage migrants and may start passing through the Delhi region from July onwards to September. The males are usually in confusing eclipse plumage at this time, and I remember being flummoxed by them at Sultanpur over a couple of seasons. When they pass through again in March, on their way back home north, they are in full breeding finery and there is no problem identifying them at this time.

Northern Pintail (Pintail), *Anas acuta*, Sand/Seekh-par

With its white-banded chocolate head, silver-grey finely vermiculated body, dazzling white breast, and stylish 'pin' feathers sticking out of its tail, the pintail (55–63 cm) is a sprightly, debonair duck, impossible to miss. Pintail are one of the most common of the migratory ducks seen around Delhi and you can get your fill of them at the Delhi zoo where they easily outnumber everything else in the duck pond. The females too are attractively scalloped in gold and dark brown.

Other areas where you can see pintails, include the Jamuna and Sultanpur. Confident, upright, and bright-eyed, with their gracefully tapering profiles, they arrive around September and leave by April. Their flight is swift and direct.

Northern Shoveller (Shoveller), *Anas clypeata*, Tidari/Tokarwala/Ghirali/Punana

Low floating, beady-eyed, and ever suspicious, the shoveller (44–52 cm) is easily identifiable by its outsized slipper-like black bill and bright plastic-orange flippers. It is a handsome duck though (if only it had better attitude!) with a glossy bottle-green or emperor-purple head, a devastatingly white breast (bustily thrust out as though to show off!), chestnut underparts and flanks, and a beautiful flame-blue forewing, embellished with an emerald speculum and white wing bar. The females are splotched in dark and light brown.

One of Delhi's most common migratory ducks, the shoveller too can be seen in fair numbers and at close quarters at the Delhi zoo, where they are second in numbers only to the pintails. Large numbers also flock to the Jamuna and Sultanpur, where they spend the balmy winter days dozing head-in-wing or drifting through the water, their spatula-like bills cleaving the surface. Like other dabbling ducks they occasionally up-end in the water while searching for tidbits and delicacies just beneath the surface, their tails pointing skywards and revolving like anti-aircraft guns looking for a target.

Shovellers arrive early (in August) and the males may still be in confusing and splotchy eclipse plumage at this time, but that broad flat bill (like an heirloom nose) and those disconcerting yellow eyes are a dead giveaway. They may stay on till as late as May or even early June.

Red-crested Pochard, *Rhodonessa rufina*, Lal-chonch/Lal-sir

Easily the most glamorous of the pochards, and a truly stunning avant-garde fellow this—and a great personal favourite. A flaming, mopsy, ginger-top head that can gleam flaxen in dazzling sunlight, a lurid vermilion bill, a black, '*bundh-gala*' jacket matched by a black rear end, pale coffee upperparts and white flanks, make up its ensemble. Rather heavy shouldered (53–7 cm), it will, however, bob jauntily on the water cork-like. The female is pale shawl brown with a darker head.

Red-crested pochards have been recorded from the Jamuna though never in large numbers. They also appear to be getting less common. Many years ago I had seen a few at Sultanpur—after a particularly good monsoon. Then, much to my delight, I came across a flock at Badkhal lake in February 1999, swimming in the company of common and tufted pochards and a bunch of coots.

Common Pochard, *Aythya ferina*, Lal-sir

A well-built though compact duck (42–9 cm) with a gingery head, charcoal-black breast and upper back and rump, and gunmetal plumage elsewhere. The female has a rufous brown head, neck, breast and upper back, and is matt grey brown elsewhere. The bill of both sexes is black at the base and tip with a broad slate-blue band in-between.

Common pochards frequent large open water-bodies, often in fairly large numbers, and extensive rafts of them can be seen on the Jamuna in winter. They are usually here by early November and are gone by the following April.

Tufted Duck (Tufted Pochard), *Aythya fuligula*, Dubaru/Ablak/Rahwara

A neat black-and-white pochard (40–7 cm) with a slicked-back '*shendi*' crest, usually recumbent but sometimes blown hilariously hither and thither. Hard golden eyes stare out at you from a sooty face, and the coal-black plumage is set off vividly by dazzling white patches on its flanks. The female is sooty black and ashy grey on the flanks.

Like the other pochards, tufted ducks like large open waterbodies and may be seen on the Jamuna at Okhla. Badkhal lake is another place to get a good look at them as sometimes they bob fairly close to the shore or to your boat. Tufted ducks arrive by early November and leave by early April. The latest I have seen them at Badkhal though was on 21 March. By 28 March they were gone.

Yellow-crowned Woodpecker (Yellow-fronted Pied Woodpecker/Marhatta Woodpecker), *Dendrocopos mahrattensis*, Katphora (applies to woodpeckers in general)

I haven't seen these small (around 18 cm or bulbul sized), profusely polka-dotted woodpeckers for a long time now, though they have always been rather unobtrusive and shy birds. Most of the body is spotted black and white, the off-white breast is barred in brown, the vent and abdomen are scarlet, the forehead pale butter yellow. The male wears a raffish scarlet crest.

Sometimes the woodpeckers draw attention to themselves by their industrious drumming—which in Delhi's cacophonous environment usually gets drowned out except in the quietest of parks and woodlands. You are likely to spot a pair, clinging on to a tree trunk in typical woodpecker (and mountaineer) style, supported by their stout tails and opposite-facing claws. The spike-like bill drums neat round holes into the wood—from which the birds pick off insects or, more seriously, which they use for nesting. Usually, three eggs are laid and both parents look after the brood.

I saw a couple of these woodpeckers several years ago at Delhi zoo: they slam-bang landed in magician-style on a dead bare branch sticking up skywards, and proceeded to corkscrew around it, hammering a brief, inquiring tatoo every now and then. I have also seen them in the woodlands of Sultanpur, quietly picking off insects from a tree trunk, with dedication and industry.

Black-rumped Flameback (Lesser Golden-backed Woodpecker), *Dinopium benghalense*, Sonera Katphora

With its ringing lunatic cackle, gleaming bullion back, and bottle-brush crest, this woodpecker is like some flamboyant extrovert. About the size of a myna (around 28 cm) it has a dull gold and black back and upper parts, a streaked off-white-greyish breast, a crimson crest and crown. The female has a stippled black and white forecrown.

Happily, this woodpecker is still fairly common in most of Delhi's parks, gardens, and woodlands. It will suddenly clamp itself onto a tree trunk with dramatic flair, then, if aware of your presence, scuttle behind it and peep coyly from around the corner so to speak. If you move your position, it will do the same, trying to ensure that it remains mostly out of sight. Occasionally, however, it becomes more interested in the creepy-crawlies that inhabit the wood and begins hammering with its great black iron-spike of a bill.

The flamebacks in some of Delhi's parks, however, appear to have embraced a more eclectic diet. One morning at the northern Ridge, I watched in astonishment as a woodpecker hitched its way down a trunk fireman-style, and hopped sideways (like a child playing hopscotch) to where bread crumbs had been scattered on some rocks nearby. It lowered its head sideways, almost to ground level and began picking up the crumbs, using its bill like a pair of tweezers. Eventually it straightened up, hop-scotched back to the base of the tree and hitched its way up again before launching itself off, undulating through the dark canopy, its back gleaming, its mad cackling 'Kyi-kyi-kyee!' laugh ringing in its wake. Apparently this has become quite a common practice and these woodpeckers are known to come down to ground level to pick up ants and other insects quite often.

Black-rumped flamebacks are known to nest virtually all through the year, especially between March and August, in the Delhi area. Again on the northern Ridge, in mid-February, I came across a bird that seemed to be putting the final touches to its home in a dead tree trunk. From inside the hole it began its carpentery, sending out a flurry of wood shavings flying out as it made the necessary modifications. The following week, however, the nest hole had been taken over by a brown-headed barbet (see following entry) and there was no sign of the woodpecker. And one week later, there was no barbet either, but just some insects humming around the entrance. They were bees; the inside wall of the tree trunk was clotted with bees, and that was that as far as the birds were concerned!

Black-rumped flamebacks are thought to pair for life, and usually three eggs are laid. The father is supposed to be an especially exemplary parent.

Brown-headed Barbet (Green Barbet), *Megalaima zeylanica*, Bada Basantha

This stout, summer-loving, leaf-green barbet is more often heard than seen—in virtually every garden and park in Delhi. About myna sized (27 cm) its plumage is mostly bark brown and leaf green; its head, neck, and upper back are brown with off-white streaks. Its lines are stubby and expression somewhat oafish. The bill is stout, yellow ochre, and adorned with a bristly moustache. The area around the eye is also yellow ochre.

Come February and the barbets begin to look forward to the delights of a Delhi summer. They stake themselves out at vantage points near the tops of trees, and their incessant metallic 'Krr-rr-kutroo-kutroo!' calls echo hollowly through parks and gardens throughout the capital. It is a sound that will continue all through the incandescent summer.

The birds themselves are difficult to spot: they are arboreal by nature and clad in leaf green, almost impossible to pick out in the peepuls, banyans, and neems they love to frequent—feeding greedily on the berries and drupes that may be on offer. Like woodpeckers, they too excavate holes in the trunks and branches of trees and, by March–April, every park with big trees will have its contingent of nesting barbets. Normally three eggs are laid and both parents share the chores.

Coppersmith Barbet (Coppersmith), *Megalaima haemacephala*, Chhota Basanta

A cute, stubby little barbet (17 cm, fat sparrow sized) with a crimson, yellow and black clown's face, large, dark soulful eyes, outsized stake-like bill (with moustache), and grass green plumage. A crimson and yellow patch on the throat and breast gives way to a charcoal-streaked yellow-green tummy. The legs are red.

Like its larger cousin, the coppersmith barbet also loves the blowtorch summer, and expresses its joy at its arrival by perching on exposed branch tops, standing on tip-toe and hiccupping 'Tok-tok-tok!' for hours together in the blazing sun. This, of course, is the establishing of a territorial claim, and often all the coppersmiths in a park or garden will be calling together to let each other know of their existence and land-holding rights.

Coppersmiths are quite common in Delhi—in parks, gardens, and woodland areas throughout the capital—being especially fond of groves of berry-laden trees, like peepul, banyan, jamun, and neem. They also excavate neat round holes in the trunks and branches of, especially soft-wooded, trees and, by February, places like the Ridge, Qudsia Gardens, Deer Park, and Delhi zoo are full of these industrious little birds hard at work. They are pugnacious little fellows too, the males not averse to (and probably enjoying) bouts of wrestling when feelings run high over territory

Plate 9 *Gulls* Black-headed gulls blizzard around the *ghats* on the river.

Plate 10 *Grey Francolin* Cheerful but wary resident of scrub country and large parks.

Plate 11 *Black Francolin* Occasionally heard, rarely seen.

Plate 12 *Northern Pintail* Handsome, debonair, migratory duck; a favourite visitor to the zoo!

Plate 13 *Indian Grey Hornbill* — Nests in the hollows of large ancient trees, very conscious of family security.

Plate 14 *Common Hoopoe* Neat and fastidious looking, but allegedly keeps a filthy home!

Plate 15 *Indian Roller Jay or Blue Jay*   Rambunctious open-country bird, getting scarcer in Delhi.

Plate 16 *White-throated Kingfisher*   Delhi's most common kingfisher; often found well away from water.

Plate 17 *Pied Kingfisher*   The riverbank and zoo are good places to look out for these dashing sky-divers.

Plate 18 *Small Green Bee-eater*   Likes open country and water; found in large parks and gardens with ponds.

Plate 19 *Blue-cheeked Bee-eater*   Monsoon visitor to open country around Delhi; the riverbank is a good place to look out for them.

Plate 20 *Great-White Pelicans* Free-flyers at the zoo.

Plate 21 *Plum-headed Parakeet* Delhi's most colourful parakeet.

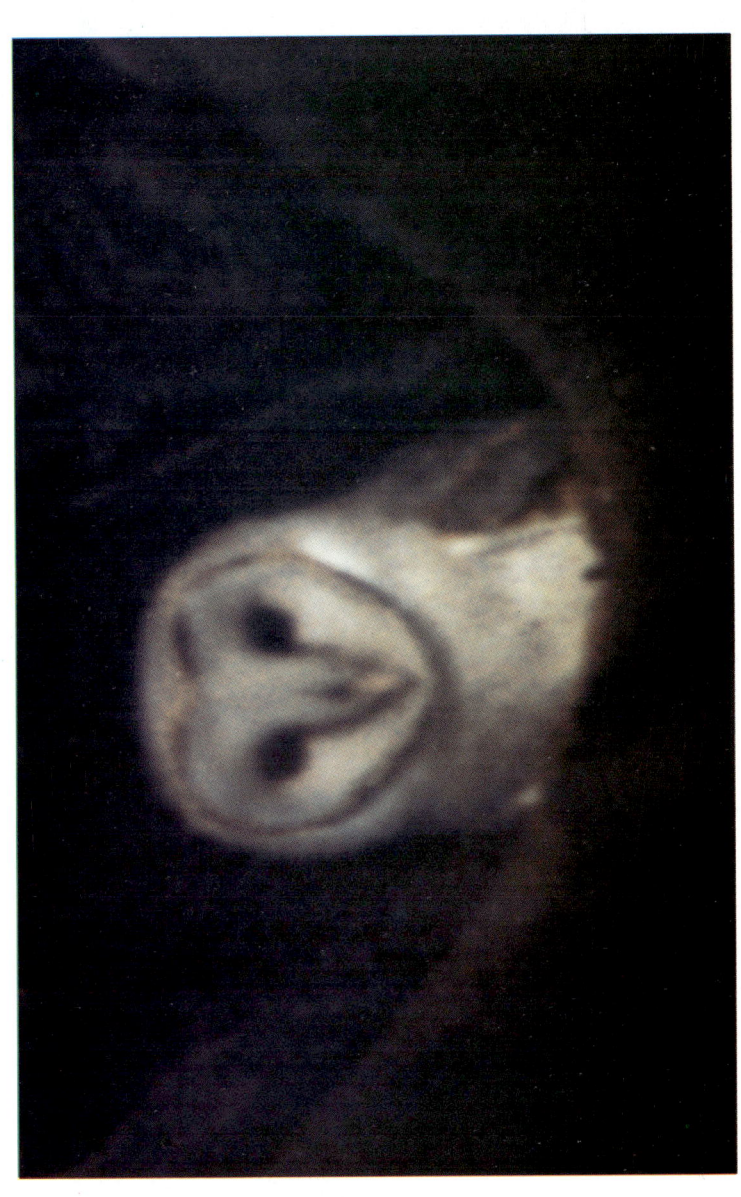

Plate 22 *Barn Owl* Rat-catcher par excellence; sadly persecuted by many.

Plate 23 *Laughing Dove* Resident in every garden; on nodding terms with everyone.

Plate 24 *Yellow-footed Green Pigeon* Usually remains hidden between leafy foliage, but gives its presence away by a strange chortling, wheezing call.

Plate 25 *Common Moorhen* Diffident inhabitant of ponds and water bodies.

Plate 26 *Common* (or *Fantail*) *Snipe*   Lurks around the edges of water bodies, usually rather wary.

Plate 27 *Blackwinged Stilt* . Almost *de rigueur* in every water body, no matter how filthy!

or females. Usually three eggs are laid, and until the fledglings attain their adult plumage they are quite unkempt and grubby looking, and don't have the flashy crimson splashes on their foreheads like their parents.

Indian Grey Hornbill (Common Grey Hornbill), *Ocyceros birostris*, Dhand/Dhamar/Chalotra/ Dhanesh/Dhanel/Lamdar

Surely a bird of many million years ago! This creaky, rusty-hinge-voiced bird, with its shabby grey plumage and huge down-curving black and off-white bill topped with its strange bony casque, looks like some antediluvian avian relic. As big (60 cm) as a black kite, it appears to be somewhat loosely put together, with a long, graduated white-tipped tail that dangles dangerously, and a flap-glide-frantic-flap-glide flight pattern that suggests its wings might come off at any moment.

But yes, grey hornbills are quaint and clumsy, and have a soft, enquiring expression in their orange-brown eyes that gives them a strange enchantment. They can be seen in nearly every large park or wooded area, and even in the tree-lined avenues of Lutyen's Delhi. They are arboreal birds that keep to the leafy upper reaches of trees, but give away their presence by their creaky, mewling 'Cheu-cheu-cheu!' calls.

They have strange old-fashioned ideas about family life too. By February they are busy courting, the male carefully selecting the choicest berries and drupes for his ragamuffin sweetheart and delicately offering them to her as though they were precious jewels. Eventually the pair start house hunting, and when a suitable hollow in a tree is located, the female enters, makes whatever modifications necessary and proceeds to seal up the entrance with her droppings. Outside, her husband does the same thing using mud, until his wife is well and truly in purdah. A narrow slit is left through which he will feed her in the following weeks until her chicks hatch.

I once watched a gentleman hornbill at Qudsia Gardens feed his incarcerated wife with berries for breakfast. He had evidently stashed away a large number of them so that he wouldn't have to fly too many sorties back and forth. Clumsily he flapped to the slit, balanced himself somewhat precariously on the trunk, using his tail as a wedge, and jerked his head up and down to manouevre a single berry to the tip of his huge bill from where his wife could delicately pick it. This procedure would be repeated again and again and occasionally the hornbill would flutter down to a more stable perch in order to regurgitate the berries without losing his balance. It reminded me of someone digging deep in his pocket and taking out chocolates one by one and nearly falling over in the attempt.

Once the chicks hatch, the female hornbill breaks out of her home and seals up the entrance again. Now both parents feed the always-hungry babies. Eventually the fledglings themselves break out of their nursery and look at the world for the first time.

Any park or garden with large old trees—especially those that have drupes or berries—is sure to attract hornbills. The New Delhi Ridge, the northern Ridge, Buddha Jayanti Park, Lodi Gardens, Deer Park, and Delhi zoo, for example, are almost sure bets for these birds.

Common Hoopoe (Hoopoe), *Upupa epops*, Hudhud

This zebra-striped, salmon-pink, lawn-loving bird is almost always mistaken for a woodpecker. With its black stippled fan-like crest—expanded when the bird is alarmed or surprised—and slender curving bill, the hoopoe (about myna sized at 31 cm) can often be found sauntering about on immaculate turf in the best addresses in town. However, it also likes open wooded country, cultivated fields, and wherever there may be juicy grubs and worms just underground.

The bird has a strange butterfly-like flight and a soft, somewhat hollow-sounding, mellow 'Hud-hud-hud!' call. Normally seen in pairs, larger gatherings can be met occasionally, though even then they remain quiet and well behaved. Hoopoes start nesting in around February, and do so in the holes of walls and trees. They are (deliberately) dreadful housekeepers, and the nest and even the babies stink to high heavens. This is just a cunning ploy to keep predators at bay and even hygiene inspectors have, up to now, obviously given them a wide berth!

Indian Roller or Blue Jay, *Coracias benghalensis*, Neelkanth/Sabzak

A well-built, bristly-faced, cinnamon-brown bird about pigeon-sized (33 cm), with stunning Oxford- and Cambridge-blue wings. The top of its head is matt blue, the neck and breast are rusty cinnamon, the abdomen pale blue. This somewhat big-headed, lumpy bird likes keeping watch from transmission lines and can truly electrify you when it unfurls its fabulous navy and powder-blue wings. Its flight is steady and deliberate.

The roller likes open, cultivated country and so can sometimes be spotted in the fields south of Okhla Barrage and such areas. Sultanpur and Badkhal are two other locations, and remember to keep your eyes glued to the transmission wires on the drive to and from these places. It is thought that rollers are now less common than they used to be in the 1970s and perhaps 1980s. March is a good month in which to keep a sharp lookout for rollers as this is when they commence their mad aerobatic courtship displays, screaming gutturally as they perform their ludicrous manoeuvres like a bunch of hoarse, love-lorn lunatics.

Rollers enjoy a diet of frogs, lizards, and large quantities of insects, which makes them valuable for farmers as pest-control experts. They nest in the hollows of trees or even in holes in walls, stuffing these cozily with rags, straw, and rubbish. Four or five eggs are laid, and the young, until their attain adulthood, look washed out and etiolated.

Common Kingfisher (Small Blue Kingfisher), *Alcedo atthis*, Chhota Kilkila/Nita/Nika Machhrala

This tiny (16–17 cm) gem-like kingfisher should alert you to its presence by its excitable, high-pitched, shrill, 'Chee-chee!' or 'Chit-it-it!' calls. Usually found near water, the bird has a shimmering sapphire-blue head, back and body, contrasted by a fiery, rust-orange breast and belly. The bill is long and dagger-like, the tail stumpy and sweet, and there is a prominent patch of white behind the eye and nape, and under the chin.

The common kingfisher also appears to have become less common in Delhi than it used to be. I used to see them quite regularly around the ponds of the northern Ridge and even the moat in Lodi Gardens back in the 1980s, but hardly ever nowadays. One place where you still might be able to see them relatively easily, is around the ponds and moats of Delhi zoo. I once spent a delightful hour watching a pair fish in the pelicans' pond. Keeping watch from an overhanging branch one of these birds caught five small silver fish in the space of 45 minutes (its partner fished from a point further away).

These kingfishers nest between March and June, in tunnels excavated in earth banks adjacent to water, and declare exclusive fishing rights over area, adjoining their homes. Five to seven eggs are laid.

White-throated Kingfisher (White-breasted Kingfisher), *Halcyon smyrnensis*, Kilkila/Kourilla

Easily the most common kingfisher of Delhi (and perhaps the whole coutry), this brilliant blue, and chocolate and white kingfisher can often be found in areas quite far away from water. With its white shirt front, chocolate 'smoking jacket' and head, brilliant turquoise plumage and scarlet dagger bill, this fellow looks like some high-class night-club proprietor as it eyes you with a wolfish grin on its face.

Its eclectic diet of frogs, lizards, insects, mice, and even small birds—in addition to fish—has made it less dependent on water than others of its tribe.

In Delhi you can see these glamorous gadfly birds in gardens and parks and woodland areas, and naturally in areas adjoining water-bodies. Transmission wires are also favoured as lookout points. Any large park, green area, or woodland—like Buddha Jayanti Park or the gardens around Humayun's Tomb—is sure to have these kingfishers in residence. The malevolent, murky ponds on the northern Ridge, for instance have perhaps two pairs of established residents: I suspect they breed here (between March and July) though I have not been able to discover where they have excavated their tunnels. Awkward-looking adolescents invariably make their appearance at the ponds during the monsoons, sitting on dead tree stumps and wagging their tails uncertainly.

White-throated kingfishers have a shrill, penetrating 'Kililili...!' territorial call, usually given from a high-up prominent perch and which is a sure indicator of their presence. Ringing cackles of laughter are usually given while arriving at, or departing from, a particular perch.

Pied Kingfisher, *Ceryle rudis*, Koryala Kilkila

A large (30 cm), striking, black-and-white kingfisher with a dramatic sky-diving fishing technique. This speckled and barred dagger-billed bird, will park itself some 10 or 15 m above the surface of a pond or water body (always facing the wind), and hover. The black eyes are sharply focused on the water, the deadly bill points straight down. The wings beat steadily, and the fanned-out tail serves as an air brake, keeping the bird rock-steady as it aims. Suddenly the bird folds its wings and dives, splashing into the water like a feathered dart. As the ripples hoop lazily outwards, it emerges, firmly holding a wriggling silver fish in its bill, and streaks off to a suitable perch where it will consume its catch. If it is not successful, it simply climbs back into the sky and parks itself up there again.

The male wears a double-stranded black gorget across his white throat, the female has a single gorget—and a broken one at that. Pairs are often encountered together.

The Delhi zoo again, is one place where you can (if you are especially lucky) watch this kingfisher perform its sky-diving stunts from fairly close up. Crows, unfortunately, sometimes play spoilsport and chase and harass the birds until they zip off, exasperated. Okhla, the banks of the Jamuna in north Delhi, Sultanpur, and Badkhal are some other places where you can see these kingfishers. The fish tanks of the Fish Farmers Development Agency at Badkhal, usually attract kingfishers (and other waterbirds) and I have seen all three species lined up there side by side, waiting with bright-eyed and ever-optimistic patience.

Pied kingfishers nest between October and April, excavating tunnels in earth banks and bringing up between five and seven young. The call is an excitable 'Chirruk-chirruk!', given on the wing.

Green Bee-eater (Small Green Bee-eater), *Merops orientalis*, Patringa

Slim, sharp-winged, pin-tailed, and masked, this elfin green bird is a familiar and welcome sight over Delhi's skies. Its clear, bell-like 'Tree-tree-tree!' trilling is usually the best clue to its presence.

Somewhat larger than a sparrow, at about 16–18 cm, this grass-green bird has a rusty-saffron shine to its head, vivid turquoise cheeks and a sleek black mask across its orange-ringed eyes. Two blunt 'pin' feathers stick out of its tail like twin antennae, and its wings, when spread, are triagular and sharply pointed.

Green bee-eaters frequent gardens and parks and especially the cultivated areas along the river front. Usually found in small groups, they like perching on transmission wires from where they launch sallies after bees, wasps, and glittering dragonflies. Sometimes they skate high in the sky, fluttering around in circles as they hawk insects, their fluting calls floating down faintly like the tinkling of far-off bells. Once caught, the prey is battered till soft, before swallowed. The birds love bathing, and once, on the northern Ridge, I watched a small group dip down swallow-like over the murky Serpentine pond, splash-dipping their breasts in the water and seemingly enjoying every moment of it.

While green bee-eaters do nest in the Delhi area—excavating tunnels in earth banks—and are residents of the city, there does appear to be some seasonal coming and going as well. I've always noticed that Delhi's skies appear to be rather more full of bee-eaters in March–April, and then again in September–October than at other times. They nest in summer—before the monsoons—and by April look their absolute best. Juvenile bee-eaters, by contrast, appear quite bedraggled and pathetic with none of the class and sophistication of their parents. One area where I think they may be nesting is in the excavated ruins within the walls of Purana Quila.

Blue-cheeked Bee-eater, *Merops persicus*, Bara Patringa

A big (around 25 cm), beautiful bee-eater with a deep contralto 'tew-tew' call, and which is a summer visitor to Delhi, arriving by April and staying on till October. Grass-green and masked, it is easily distinguished from the green bee-eater by its larger size, pale salmon-chestnut throat patch, yellow chin, white and turquoise eyebrow (supercillium), and blue-green ear patch.

In Delhi, I have admired them in the fields near the cremation ground on the east bank of the Jamuna opposite Mansarovar Park in June. Both juveniles and adults have been seen regularly at Sultanpur in May and June, and in the past they have reported to be breeding in the low sand hills in the areas adjoining the jheel. Usually five eggs are laid. Dragonflies and cabbage-white butterflies are supposed to be their favourite food.

Pied Cuckoo (Pied Crested Cuckoo), *Clamator jacobinus*, Kala Papiha

A big (33 cm), swashbuckling, black and white cuckoo this, with a rakish gelled crest, ringing, metallic 'Le-yew! Le-yew!' call, and a bit of mystique attached to it. Nearly always you will hear it before seeing it, usually around mid-or end-June, and the somewhat wanton call somehow always raises your dehydrated mid-summer spirits. Keep a sharp lookout for a tall, handsome, debonair cuckoo, black on top from head to tail, pure white underneath, and with white band on black wings and white-tipped elongated tail. Rather like an outsized millionaire bulbul in coat and tails, all set for the races!

They say these crazy *bon vivants* ride the monsoons winds over the Arabian Sea, all the way from East Africa to spend the monsoons romancing with us, but no one really knows. (They are resident birds in the south.) What we do know, is that almost always the monsoons break in Delhi very soon after these wanton spirits have been heard and sighted. And they come to Delhi (and the north of the country) for romance, courtship, and dalliance. Males conduct crest-raising skirmishes with other males, and then chase the girls around with shameless abandon even in such sanctimonious (and bureaucrat-ridden) places as Lodi Gardens and Buddha Jayanti Park. The New Delhi Ridge is another favourite haunt for their peccadillos. And there's worse to come: if the female gets into trouble, she simply slips into the nearest jungle babbler's nest and divests herself of her unwanted burden. And then it's back to fun and games! You can enjoy the company of these madcap romantics between June and September.

Common Hawk-cuckoo or Brainfever Bird, *Hierococcyx varius*, Papiha/Kapak/Upak

Better known as the brainfever bird, this is one creature that can drive you to dementia if you go looking for it. From March onwards it will begin its taunting, infuriating call, screaming out 'Brain-fever! Brain-fever! Brain-fever!' ad nauseum as you go stomping about in the heat, looking for it. The hotter it gets, the more maddening the calls become.

Slightly larger than a pigeon (around 33 cm) and with a longer tail, the common hawk-cuckoo is greyish brown on top, off-white and ashy-rufous below with a rusty weave across its abdomen. It looks like it's trying to pass off as a shikra. However, it has a weak pigeon's bill, and mad staring eyes that suggest only the terminal stages of lunacy. Its manner of flight too is fluttery and panicky, with none of the deadly purpose of that of the hawk it tries to mimic.

This lunatic creature is a summer visitor to Delhi, occupying the leafy reaches of large parks and gardens between March and October. Lodi Gardens, the gardens around Humayun's Tomb, the New Delhi, and northern Ridges are some favourite haunts. Thankfully, perhaps, this neurotic bird does not bring up its own young—eggs are disposed off in the nests of ever-accommodating and foolish jungle babblers (not that it makes much of a difference though—the young end up being just like their biological parents!). Partial to hairy caterpillars, insects, and berries, the brainfever bird usually remains up in the leafy treetops, and perches in a strange upright manner that suggests an imminent panic attack. The birds may remain in the Delhi area throughout the year: I have seen them on more than one occasion at Sultanpur in winter. Clearly out of their element in this season, the birds sat glumly out in the open and uttered not a sound, allowing me to come close enough to jeer at them! Every dog has his day!

Asian Koel, *Eudynamys scolopacea*, Koel

The koel is another mad monsoon romantic, and an abundant one at that. The male (43 cm) is jet black, glossed with a metallic green sheen, has staring ruby eyes, and an effeminately narrow waist. His wife is bark-brown, profusely polka-dotted and barred with white, and perfectly camouflaged for her sorcery.

Koels start their incessant 'Kuoo-kuoo-kuoo!' calling—in parks, gardens, and just about everywhere—by March–April, and by August they are vocally ballistic, bubbling over with hysterical excitement. Though difficult to observe because they are so arboreal, they are worth seeking out for their soap-operatic shenanigans.

At first there is much racing and chasing through the trees as males pursue each other over territorial matters, and females chase rival females. It's almost as though some medieval family vendetta is being played out up there amongst the valentine-leafed peepuls. Then the males chase the females in courtship and, after this dizzy dalliance is done with, it is time for trickery. The male flutters close to a crow's nest and is duly chased by the outraged birds, leading them away with a fine song and dance while his wife, dappled and speckled and virtually invisible, slips in and deposits her eggs into the unguarded crows' nest. (One silly crow pair allowed thirteen koel eggs to be laid in their nest—probably by more than one koel female.)

On one gloomy monsoon afternoon a pair of koels suddenly arrived on the bottlebrush tree outside my window. The male called belligerently and shuffled closer to his wife (or girlfriend) who looked a bit nervous. Then another male flew out from a nearby tree and plonked himself between the female and the first male. Immediately a *tu-tu-main-main* exchange—or rather, a yodelling match broke out, though each male would wait for his turn to shout: the idea seemed to be to out-shout the rival, not yell at the same time. Completely unnerved, the female flew off, and now the two boys hitched themselves defiantly closer to one another, glaring at each other with their terrible blood-globule eyes and yelling for all they were worth. Eventually the intruder called it off and flew away (in pursuit of the female, I wonder?) and the victor was left all alone on his tree.

Koels appreciate fruit and berries as well as insects and caterpillars.

Greater Coucal (Crow Pheasant), *Centropus sinensis*, Coucal/Mahoka

A whopping great (48 cm), coal-black, crow-like bird with rich russet wings, a long, broad, black graduated tail and bulging blood-globule eyes. An undergrowth skulker and shuffler, it is usually hard to spot, but gives itself away by its hollow echoing 'Coup-coup-coup!' calls that can make you quicken your steps in shadowy Delhi parks in the evening! But if you do get to see one from close enough, you may notice that it even has eyelashes.

Coucals are found in parks and gardens where there is plenty of undergrowth and shrubbery, like the New Delhi Ridge for example. Open cultivated areas—near Okhla for example—and reed beds adjoining the Jamuna are also favoured. If you drive south along the road on the western side of Okhla Barrage, you may be able to spot many fat, shiny specimens sunning themselves from bush tops as the sun comes up.

The birds belong to the cuckoo family and are inveterate baby-bird connoisseurs. Unlike other cuckoos, however, they build their own nests—large, domed structures of grass and straw and twigs—and commence courting by April. Three or four eggs are laid.

Alexandrine Parakeet (Large Indian Parakeet), *Psittacula eupatria*, Hiraman Tota/Rai-tota

A big (53 cm), macho grass-green parakeet, with a powerful red nutcracker of a bill. A hoarse, guttural—'Keeak!' or 'Kee-ah!—call and a relaxed leisurely manner of flying. A black moustachial streak, attractive rose-pink collar (gentlemen only), and maroon shoulder patches complete the ensemble. More dignified than the madcap, rose-ringed parakeet (see following entry) even though its domed head and bulging eyes give it a somewhat wild, lunatic expression.

Alexandrine parakeets are nowhere as common as rose-ringed parakeets, though some favourite haunts are Qudsia Gardens, the gardens around Humayun's Tomb, University Gardens, and Lodi Gardens. They breed between December and May in the hollows of large old trees (in University Gardens for example) and in holes in the walls of old monuments and buildings. Two to five eggs are laid. They are a favourite amongst the *mithoo-tota* pet bird brigade, though it is always so much nicer to watch a couple of these hefty parakeets necking ardently high up in the banyans of Qudsia Gardens, than turning round and round in and squawking in frustration in tiny cramped cages.

Rose-ringed Parakeet, *Psittacula krameri*, Tota

Surely this parakeet should be a frontrunner for the position of Delhi's state bird! Slim, compact (42 cm), chilly-green, with a curved red bill and rose-pink collar (only the males), these garrulous, exuberant 'Punjabi' parakeets can be found screaming across Delhi's skies in gangs of a dozen birds or more, especially in the mornings and evenings. Of course, when they arrive on the guava or mango or bottlebrush tree in your garden, complete silence is maintained. It is astonishing how many of these '*hari-mirch*' birds you can flush from a single tree.

Rose-ringed parakeets may be common but they're always entertaining to watch. Especially while breakfasting on breadcrumbs and birdseed scattered for them (and others) in parks every morning. Buddha Jayanti Park and University Gardens are two such breakfast parlours, with scores of parakeets in attendance every morning.

The birds, which are abundant all over Delhi, nest between January and May, in the hollows of trees and holes in the walls of buildings and ruins. Four to six eggs are laid. The courtship display of the male is quite ludicrous and something that must be watched at every opportunity. He sidles up to his love, sideways on, one claw raised reassuringly and then is suddenly smooching her passionately smack on the beak (eyes almost rolling out of his head) before backing off as though contrite, only to approach her again from the other side!

Rose-ringed parakeets (and others of their clan) are trapped in large numbers for the pet trade, so if you do spot a nesting hollow, it is better not to draw attention to it. There are just too many vagabonds and little boys around, ever ready to shin up and pinch the chicks. Some people buy trapped parakeets in large numbers and then piously 'free' them in parks or the Ridge, in the hope that the Gods will smile benignly on them and forgive their latest transgressions. All they accomplish is to ensure that still more parakeets are caught from the wild and offered for sale once again. One in every three birds so caught dies before coming on to the market, so this is actually quite a murderous business.

Plum-headed Parakeet (Blossom-headed Parakeet), *Psittacula cyanocephala*, Tuyia Tota

The least common of Delhi's parakeets, the tuyia is also the smallest (36 cm) and most attractive. Males have a rich plum-coloured head and bluish collar; females have a cloud-grey head and yellow collar. The bill is papaya orange. The plumage is green, and the two central feathers of the tail are blue green tipped with white.

Tuyias streak across the skies, uttering characteristic pleasant-sounding, interrogative 'Tooi? Tooi?' calls in their wake. Not too common in Delhi, they can, however, nearly always be found in University Gardens every morning, where they join the regular rose-ringed parakeets for breakfast. Qudsia Gardens and Buddha Jayanti Park are other likely locations where you can see these birds, as is the area around the Badkhal Lake complex. The birds breed between December and June, nesting in hollows of trees and holes in walls. Four to six eggs are laid.

House Swift (Little Swift), *Apus affinus*, Ababeel/Babeela

A sleek, sooty, fastback-winged little bird (15 cm) with a broad white band across its rump and a squared-off tail. Nearly always seen zipping in and out and about old monuments and ruins, or scissoring high in the skies in great ball-like flocks. Humayun's Tomb, for example, seems to be one favourite area, while others would include the monuments on the the Lodi Golf Links, the tombs in Lodi Gardens, and Purana Quila.

The birds flutter and glide, dip, dive, and swerve deftly through the air (and in and out of narrow gunslit windows and doorways) emitting high-pitched, squeaky 'Siksiksik-sik-sik-siksiksik!' twitters as they whiz around, hawking tiny insects, beaks agape.

House swifts nest between February and September. Their nests are little scruffy balls of feathers and straw stuck together by saliva and cemented to the angles between walls and ceilings and similar suitable locations. Two to four eggs are laid. The birds are thought to be local migrants to Delhi, leaving the capital during winter.

Barn Owl, *Tyto alba*, Kuraia/Kurail

A beautiful pale-alabaster and mushroom-gold owl, finely stippled with black and white, with a grave heart-shaped face frilled with a ruff, and creamy white underparts spotted with dark brown.

In the past, the barn owl has been described as being a scarce resident of Delhi, though I have in recent times been hearing several delightful stories of barn owls giving people a fright in some of east and south Delhi's mass housing colonies, appearing ghost-like at windows and on balconies after dark. The ancient gnarled trees in Qudsia Gardens have been home for these lovely birds for many generations now. But a walk in the gardens at night, when they are in attendance is only recommended for the stouthearted; their cackles and screeches can make your blood run cold.

Barn owls usually live close to man and will gladly nest in the ruins of old buildings and monuments. They are believed to breed throughout the year—in winter certainly—and bring up four to seven young. Perhaps one reason why they are seen so infrequently is because they are purely nocturnal by nature, and vanish into their hollows at sunrise usually being chivvied away by crows, babblers, and other birds. (Most other birds seem to have a long-standing vendetta against owls.) Barn owls are unfortunately sought out for their body parts, which are believed to possess medicinal properties and are also used in occult practices. It is therefore prudent not to draw attention to the presence of these birds, or their nesting sites, when you come across these. The birds (as are all owls) are considered to be creatures of ill omen, which is ridiculous, because they are actually, the best, most silent rat-catchers in the avian kingdom. And, certainly, Delhi has far too many rats!

Spotted Owlet, *Athene brama*, Ulloo/Kukushat

A rotund grey-brown owl (21 cm), profusely spotted in white, with great golden eyes and a rather big round head. This interesting little owlet is found in parks and gardens throughout Delhi and can occasionally even be spotted during the day, dozing up in the leafy reaches of some tree. They nest in the hollows of trees—between November and April—raising three or four young.

Crepuscular and nocturnal, they emerge at dusk, uttering harsh, bad-tempered sounding 'Chrr! Chrr! Cheevak!' calls as they sort out hunting blocks for the night. Beetles, other insects, mice, and young birds are prey.

Look hard up in the big neem trees in Lodi Gardens, for example, and you are almost sure to spy a spotted owlet or two, staring down back at you out of those circular golden eyes. Other places where I have regularly met these delightful birds include Nicholson Cemetery, Qudsia Gardens, University Gardens, and Sultanpur. The trees lining the avenues of Lutyens' Delhi are sure to have a good number of these birds in residence too.

Baby owlets have a delightful way of bobbing their heads—like Bharatnatyam dancers—while staring you down, in an attempt to intimidate you and to make you go away. Like other owls, spotted owlets are not at all popular with other birds, and are mobbed and chased away by them at every opportunity.

Rock Pigeon (Blue Rock Pigeon), *Columba livia*, Kabutar

It is difficult to believe that this plump, waddling, oversexed, slate-grey pigeon with its shimmering, iridescent throat was once (and some still are) a resident of wild rocky gorges and terrifying cliff faces. The species has moved into cities big time, roosting and nesting just about everywhere—in air-conditioning ducts, service station bays, railway sheds, and usually where they cause the maximum inconvenience. Despite this, they are well loved and well fed by residents and there are many 'pigeon only' feeding spots scattered throughout Delhi. Large flocks spin around the skyline of the city, usually panicked into the air by a tyre burst or backfiring engine or crackers. They roost in trees in gardens and parks, or on the ledges of buildings and skyscrapers. The birds love bathing and often fly down to freshly watered lawns to bathe in the company of sparrows and mynas. Otherwise, they're normally to be found seducing one another.

Laughing Dove (Little Brown Dove), *Streptopelia senegalensis*, Chhota Fakhta

A compact, endearing little dove (25 cm), sandstone brown, with a lilac-grey head, neck and breast, and a sandstone and black miniature 'chessboard' pattern on either side of its neck. The tail is rather long and gives the impression of being dragged along behind the bird as it waddles forth.

Commonly encoutered in parks and gardens, often entering verandahs and balconies, the laughing dove strolls around somnambulistically, uttering its dozy, self-satisfied, 'Coo-roo-roo-roo!', calls, as though trying to talk itself to sleep. The birds breed between March and October, building flimsy and untidy nests out of thin twigs (with dangerous see-through flooring through which eggs invariably fall). Normally, two eggs are laid.

Eurasian Collared Dove (Indian Ring Dove), *Streptopelia decaocto*, Dhor Fakhta

A sandy-fawn, pigeon-sized (33 cm) dove with a black half-collar on the back of the neck and a somewhat overpowered look. Quite common in Delhi, collared doves can be found in parks and

Plate 28 *Red-wattled Lapwing*   The 'did-ye-do-it?' accuser.

Plate 29 *Darter* Stunning but not as common as desirable.

Plate 30 *Indian Cormorant* The most convenient place to see breeding colonies is the zoo.

Plate 31 *Egrets* In fine lace and satin, but with a city dweller's hard eyes. Inset (left): Little and Intermediate Egret. (right): Cattle Egret — another common resident.

gardens and cultivated areas everywhere in the capital. They often join breakfasting parties of parakeets, pigeons, and sparrows at the feeding spots in parks every morning, bobbing their heads up and down rapidly as they pick up scattered corn or grain. Buddha Jayanti Park is one such favourite adda.

The birds breed between March and October, though the males seem to perform their typical territorial flights all round the year. From a high perch or branch, they take off almost vertically, climbing steeply into the sky and then plane down to another perch or branch in a wide arc, wings spread taut, tail fanned out prettily, uttering a guttural 'Koo-koo-kroo!', as they descend. While the male assiduously collects thin twigs for the nest, the female assembles these into the clan's typical flimsy platform-like structure—not the safest place for her eggs.

Nesting birds are often harassed by crows, but the female will valiantly and determinedly chase away the villains. Collared doves also enjoy scrapping with one another, and combatants will fly at each other hammer and tongs during these duels. Courting males puff up their throats and waddle after the females, bowing and ducking in the usual besotted manner of their clan.

Red Collared Dove (Red Turtle Dove), *Streptopelia tranquebarica*, Seroti Fakhta

The only species of dove in which the sexes dress differently: the male is an attractive sandstone-pink fellow (about 23 cm) with a slate grey head, black half-collar, and dark liquid eyes. His wife looks very like a collared dove but is much smaller and somewhat darker, and is usually seen in the company of her husband. The birds have a distinctive masculine-sounding 'Groo-gurr-goo!' call, repeated rather rapidly.

Red collared doves are not very common in Delhi and are found sparingly in parks and gardens, woodland areas, and open cultivation. Rather shy and retiring, they seem to prefer staying up in the treetops, though they often give away their presence by their guttural calling.

I have seen these doves quite regularly in University Gardens—though never in any great number. However, a fairly large number of birds would arrive regularly in the woodlands around Sultanpur in May and June and settle down to nest in the acacias there. After the monsoons, most of the birds would be gone, so it does appear that these doves have somewhat gypsy lifestyles. The nest is the usual flimsy platform of twigs. The breeding season is thought to last between April and October, and two eggs are usually laid.

Yellow-footed Green Pigeon (Green Pigeon), *Treron phoenicoptera*, Harial

A large (33 cm), plump, yellowish-green pigeon with an ashy-grey head and neck, lilac patch on the shoulder, and chrome yellow legs and feet. It is usually very difficult to spot in the leafy reaches of the neems and peepuls it usually favours, but is equally conspicuous early in the morning—or in the evening—when it likes to sun itself from high-up branches, twigging its tail up and down slowly and thoughtfully like a water diviner.

Green pigeons love figs and other berries and will clamber about the branches parakeet style in search of the fruit. They are not commonly seen in Delhi, more because of their superb camouflage than any other reason. I have, at one time or another, seen them in every large park or green area that I have visited: Qudsia Gardens, University Gardens, Buddha Jayanti Park, the northern Ridge, the trees in the grounds of Delhi Gymkhana Club, and at Sultanpur, to name a few places.

The birds have a curious, wheezy, whistling-chortling call, a sound impossible to reproduce on paper but unforgettable once heard. Green pigeons nest between March and July and usually two eggs are laid in the trademark flimsy nest.

Sarus Crane, *Grus antigone*, Sarus

India's only resident crane, this tall (156 cm), stately, cloud-grey bird has a scarlet head and upper neck (and ashy crown), orange eyes, black-tipped wings, and a large, long bill. Probably less common now than in earlier years (till the 1980s), the birds can usually be seen in pairs (they marry for life) in flooded fields and pastures. They have bred in Sultanpur National Park for many years and the jheel and surrounding fields remain good places for sighting them.

The best time of the year to see them is just around the monsoons, in July and August, because they really are the most besotted of romantics at this time of the year. They will toss back their heads and bugle at the glowering thunderheads, they will hop, skip and jump and fluff up their skirt tails and wings in a marvellously cabaret-dancer-like manner, and will infuse you with a *joie de vivre* in a way that very few other birds can. Their nest, comprising a huge mound of reeds and rushes piled on the ground, is usually located on a raised area in the middle of a paddy field or flooded area. Both parents take great care of the young— usually one, sometimes two—and by around December or January you can sometimes see a family out for a sedate stroll in the fields. The youngster, with its blondish head and cinnamon plumage, is usually carefully ushered between its watchful parents.

Sarus cranes fly strongly, and usually bugle to one another as they do, though sometimes it can be difficult to pinpoint their haunting calls. The birds eat grain and shoots as well as insects and reptiles. While they are revered in India because of their fidelity, the rampant use of chemical insecticides and fertilizers is thought to be responsible for the fall in population of these dignified yet dotty birds.

Delhi zoo is another place where you can see them at close quarters, even though these may have their flight feathers pulled out.

Common Crane, *Grus grus*, Kulang/Kurunch

Along with the Demoiselle crane, this migratory crane spends the winter raiding crops and feigning innocence in the riverine tracts and jheels around Delhi. About 110–20 cm tall, these greyish birds have a sooty-black head and foreneck, with a white stripe behind the eye running down the sides of the neck and a patch of red on the crown. Thought to be less common in the Delhi area than the deomoiselles, they were regulars at Sultanpur in winter, where (along with the others), large flocks numbering 500–600 birds would arrive every evening after having spent the day despoiling the surrounding fields. They would take off before dawn the following morning. It was always a pleasure to listen to their echoing, evocative 'Krooah! Krooah!' bugling calls as they sailed past, in great wavy echelons or arrowhead 'V's spread across the brandy-gold evening sky, and to catch the pre-dawn excitement of their mass take-offs the following morning.

Demoiselle Crane, *Grus virgo*, Karkarra

The smallest (90–100 cm) of the cranes, and somewhat stocky and flouncy in appearance. The demoiselle crane is grey, and wears elaborate down-curving tufts of white feathers behind the eyes, and a flouncy if ragged-looking 'shawl' of drooping black feathers over its breast. The tail feathers too (as also on the other cranes), droop and curl over its bottom.

Once regarded as 'uncommon' in the Delhi area, these migratory cranes have flocked to Sultanpur in recent years in large numbers, obviously enjoying the winter harvests of the farms nearby. Their call is a somwhat high-pitched, far-reaching 'Garoo...garoo...!' that can give you goose—or rather garoo—bumps on clear, cold, full-moon winter nights!

White-breasted Waterhen, *Amaurornis phoenicurus*, Dawak/Dahak/Dauk

This stumpy, big-footed marsh bird (32 cm) is slate grey (and tinted bronze green), with a chalk-white face and breast and a ginger patch beneath the tail, which is often flicked up to display this. The bird is common in water-bodies all over Delhi—at Okhla and the ponds on the Ridge, the zoo, and Buddha Jayanti Park—to name just a few places.

Usually you can spot it—singly or in pairs—humping cautiously along the edge of the water, picking up insects, molluscs, grain, and plant matter, flicking its tail up and down as it proceeds.

Waterhens breed in the monsoons, between July and October, and at this time can startle the wits out of you by the stentorian 'Krr...kwak...kwak...kwak!' calls they produce, especially on gloomy, overcast days—and sometimes all through the night. They can also be spotted standing alertly high up on the branches of trees, surveying the area around them like prospective property developers.

I know that the birds breed regularly on the northern Ridge, as, after every monsoon, there are broods of fuzzy black chicks following their parents about, or scuttling under their wings for shelter. The birds are regular visitors to the feeding spots too, some of which are quite far away from the water.

Purple Swamphen (Purple Moorhen), *Porphyrio porphyrio*, Kaim/Kalim

A big, balloony (43 cm) purplish-blue rail, with an outsized pillar-box-red bill and frontal shield, enormous dull red legs and feet, and what appears to be a ridiculous white handkerchief tucked under its tail, which is constantly being flicked up and down. This enchanting if clumsy-looking bird can be most easily seen stomping about on the rafts of water hyacinth on the Jamuna—wherever these conglomerate. I have also seen the birds at Sultanpur and Badkhal Lake.

Purple swamphens are monsoon breeders and usually quiet, keeping in touch with each other with a soft, chucking, 'Chuk-chuk-chuk!'. Their song has been described as an extended rattling, nasal, 'quinquinkurr...!'. The birds are residents and live off shoots and vegetable matter. They breed during the monsoons, building large pads of interwoven reed flags and rushes on rafts of vegetation. Between three and seven young are raised.

Common Moorhen (Indian Moorhen), *Gallinula chloropus*, Jal Murghi

A compact (32 cm) sooty-brown, large-legged water-bird, with a black head and neck, dark bronze upper plumage, and slate-grey underparts. A yellow-tipped red bill, white flash across the flanks, 'handkerchief' under the tail, and yellow-green legs (red above the knee!), are prominent pointers to its identity.

Like white-breasted waterhens, moorhens are common around the water-bodies of Delhi and may be found in much the same places, though may be less frequently seen because they are more shy and perhaps less abundant. They are both resident as well as migratory, and breed during the monsoons. A good place to see them is in the ponds of Delhi zoo, where small parties of eight or nine birds may be seen together.

Common Coot (Coot), *Fulica atra*, Dasari/Dasarni

This hen-sized (36–8 cm) water-bird is a winter migrant and sometimes misidentified as a duck of some kind. Sooty black from head to tail, it has a chalk-white frontal shield, pale lilac-white bill, and dull green (and big) legs. Large flocks of these birds can be seen freckling the Jamuna at Okhla, often in the company of big rafts of duck. I have seen them consorting with pochards at Badkhal Lake and also, with dabbling ducks such as shovellers and pintails at Sultanpur. The birds arrive by September and depart by May. I have often watched them being harassed by marsh harriers at Sultanpur. The birds eat mainly aquatic vegetation. When alarmed they tend to 'run' across the water on their big feet, flapping their wings frantically, and creating a great splashy uproar that sounds like some great ogre vigorously rinsing out his mouth!

Common (or Fantail) Snipe, *Gallinago gallinago*, Chaha

Snipe are small (this one is just 27 cm), somewhat delicate-looking marsh birds with relatively long poker bills (5 cm), and are cryptically but beautifully striped and patterned in buff, dark brown and black. Usually difficult to spot in the marshy, soggy areas they frequent, they often lie doggo until nearly stepped upon. They zing off, in a low, swift, zig-zag manner, their bills tending to point downwards even as they flee.

The common or fantail snipe is dark brown above, with streaks of buff, black, and brown across its plumage. The head is prominently striped in buff and dark brown. The very similar, Jack or pintail snipe (*Lymnocryptes minimus*) is a smaller version of the above (just 21 cm) and is slower and more level in flight, but almost impossible to tell apart in the field.

Snipe are winter visitors, arriving by September and departing by April. I have, on several occasions, nearly stepped on them at Sultanpur and have yet to spot one before it spots me.

Black-tailed Godwit, *Limosa limosa*, Gudera

Godwits are usually unmistakable because of their enormously long Pinocchio bills. The black-tailed godwit is a largish (36–44 cm) mottled-brown and white wader, with a disproportionately long slender bill, orange at the base, and slightly upcurved. A white bar spanning the trailing edge of the wings and a broad black band at the end of the white tail will help you identify it as it flees from the scene. Godwits are migratory and fairly common. I have seen them occasionally at Sultanpur.

Common Redshank (Redshank), *Tringa totanus*, Chota Batan

A moderately tall (27–9 cm) 'salt-and-pepper' wader, with a medium-sized, straight, black and red bill and orange-red legs. The upper parts are a mottled grey brown, the greyish-white breast is streaked, and when fleeing, a broad white trailing edge to the wings, and white rump and lower back are prominent. Redshanks have a clear double noted 'Tiweet-tiweet!' whistle, and when suspicious will bob their heads and then tails up and down.

Most redshanks in the Delhi area are winter visitors, though they have been occasionally seen even in summer. They breed in the Himalayas. Pairs or small parties may be seen on the banks and mudflats of the Jamuna—at Okhla and elsewhere—and I have met them regularly at Sultanpur. They are wary and do not tarry too long for inquisitive birders.

A slightly taller—and not as common—relative is the **Dusky** or **Spotted Redshank** (*Tringa erythropus*, Gatan) which, overall, is of a darker, more sooty and spotted complexion, and lacks the white wing bar of the common redshank. While its bill is also red and black, it is longer and more needle-like.

Unfortunately most waders dress in drab, nondescript, and maddeningly similar salt-and-pepper casuals at this time of the year and can be difficult to identify—especially as they are normally seen from long distances. They begin to change into more distinct breeding attire by around March and April, just before they take off north again.

Marsh Sandpiper, *Tringa stagnatilis*, Chota Gotra

A dainty (25 cm) grey-brown wader with a pure white supercillium, forehead, sides of head, rump, and lower back. The sides of the breast are flecked in brown, the bill is black, the legs slender and dark green. A winter visitor—thought to be a passage migrant—the marsh sandpiper, like others of the clan, can occasionally be spotted on the banks of the Jamuna and at jheels like Sultanpur. Its call is a piping 'Che-wheep, che-wheep!', usually given when the bird takes off.

Green Sandpiper, *Tringa ochropus*, Sobuj Gotra (Bengali)

A squat dark greeny-bronze sandpiper (24 cm) with a white tummy and vent. In flight, the dark undersides of its wings contrast with its flashy white rump. It flies fast and low, with a shrill urgent 'Chwee-chwee!' call trailing in its wake.

Green sandpipers are fairly common winter visitors and can be spotted on the banks of the Jamuna, in canals and nallahs (even filthy ones) flowing into the river, and at jheels like Sultanpur. They usually arrive by September and leave by April.

Wood Sandpiper, *Tringa glareola*, Chupka

A compact (18–21 cm) grey-brown sandpiper, delicately polka-dotted with white. A prominent white supercillium and white rump and underwings are other pointers to its identity.

Thought to be the commonest sandpiper in Delhi, small parties may be seen in the company of other waders, working industriously about the mudflats of the river, the banks of canals, flooded areas, and jheels like Sultanpur. When flushed, the bird blurrs off with an alarmed 'Chif-if-if!' call, its white rump contrasting with the barred tail. The birds have been recorded arriving as early as July and staying on till May. As summer approaches they get darker, and their icing-sugar spots begin showing up beautifully.

Common Sandpiper, *Actitis hypoleucos*, No local name

A small (21 cm) olive-brown (above) and off-white (below) sandpiper, with a thin white supercillium, white wing bar, and dusky rump. Like the others, a winter visitor, arriving by the end of July and staying on till the end of April.

Common sandpipers can be spotted dipping into the mud and squelching at the edge of the Jamuna, or pools and tanks, either singly or, occasionally, in twos and threes. I have also seen them at the ponds at Delhi zoo. They fly with rapid wing beats interspersed by brief glides, their wings arched stiffly downwards and almost cleaving the water. The call is a shrill 'Twee-we-we!'. The birds habitually bob their heads and tails up and down in a rather amusing all-knowing manner.

Little Stint, *Calidris minuta*, Chota Panlowwa

A dumpy little runabout (just 14.5 cm) in mottled grey brown (above) and white (below), with blackish legs and bill. The smallest of our waders, it is a winter, passage migrant and may be commonly seen scuttling about busily on the muddy riverbanks and at the edges of jheels such as Sultanpur. Little stints have been recorded arriving in late July and they return north in April, by which time many have acquired their lovely rusty breeding plumage. Some of these little visitors may have come from as far away as Siberia.

Ruff and Reeve, *Philomachus pugnax*, Gehwala

A strangely named wader where the sexes have different names: the males (standing about 26–32 cm) are known as ruffs and the females (20–5 cm) as reeves. These grey-brown birds are boldly scalloped and streaked on their upper parts and have orange-yellow legs and short black bills.

They are winter, passage migrants, and in recent times have demonstrated the importance of preserving wetland habitats like Sultanpur. At the beginning (around September–October) and end (March–April) of every migratory season, more than 5000 ruff and reeve would swirl like dark swathes of smoke in the skies above Sultanpur, and then settle in the shallows of the jheel looking like vast expanses of clotted clay. They would disappear—probably further south as the winter took hold, but return as spring approached. The birds remained faithful to Sultanpur even as the lake began drying out—though they stopped coming once the water was completely gone.

The birds have a rather correct, upright posture and, at the time of departure, some of the males begin to turn chestnut around the throat prior to acquiring their famous Elizabethan breeding dresses (wherein they wear elaborate, richly coloured ruffs around their necks).

Greater Painted Snipe (Painted Snipe), *Rostratula benghalensis*, Rajchacha

A busty, balloony bronze-green wader (males just 25.5 cm, females 28 cm) with a rich chestnut head, neck and breast, white undersides, and long bill, slightly down-curved at the tip. Most prominent, however, are the cream 'spectacles' and 'braces' (or knapsack straps) worn by the female, who is more strikingly attired than her husband. The male is slightly smaller and somewhat faded by comparison and doesn't sport the rich chestnut colours on the throat and breast.

Painted snipe are resident birds though not very common and difficult to spot (because they sit very tight indeed in thick cover) but are well worth looking out for. Scan the reeds and rushes on the edges of the Jamuna, or the canals flowing into it, or marshy patches and jheels, and you may be pleasantly surprised. I once saw two plump pairs sitting in a reedy clump at the edge of the Jamuna (south of Okhla) and pairs crouched low in the grassy edges of Sultanpur jheel. The birds are wary and do not tolerate close approach. Their flight is somewhat weak and fluttery, but their ability to dive into cover and vanish is exemplary.

Painted snipe are polyandrous; females will fight other females for the right male, then lay its eggs (snipe nest practically all through the year) and leave him to hatch them and bring up the family while she goes looking for another male (this is one way she can ensure having the maximum number of babies; it's time other birds caught on to this trick!) The birds are crepuscular and feed on paddy grain, insects, vegetable matter, molluscs, and the like. The call has been described as the soft, hollow 'oook' you hear when you blow gently into a bottle.

Eurasian Thick-knee (Stone Curlew), *Burhinus oedicnemus*, Karwanak/Barsiri

A tall (40–4 cm), upright, 'standing-to-attention' sandy-brown wader, marked in dark brown, with swollen knees, yellow legs, a big head, and great, globular golden eyes. Not very common and difficult to spot, one of the best places to get a good, relatively close look at these birds is at the Delhi zoo, where they seem to like keeping silent vigil on the 'island' in the main duck pond. I've seen over a dozen birds there at the same time, and it is amazing how, as you keep looking, their numbers seem to increase magically. On hot days they like crouching in the shade of the scrub and may squat down, almost merging invisibly into the ground. I've seen the birds at Sultanpur too—but from a much greater distance.

Thick-knees are resident birds and breed between February and August, laying two superbly camouflaged eggs on the ground or in a rough scrape. In the past, eggs and young have been found in the dry bed of the Jamuna. The birds are relatively inactive during the day, and come to life at dusk, when they go about picking up insects, worms, and small reptiles. The call has been described as a series of clear, whistling screams, 'Pick, pick, pick, pick...wick, pick-wick, pick-wick!'.

Black-winged Stilt, *Himantopus himantopus*, Goz Paon

This tall (38 cm), slender black and white wader, perched on 25-cm-long, slim pink legs and equipped with a sharp, skewer bill is a common sight wading in the shallows of the Jamuna or ponds and pools around Delhi. The male is black (with tints of bronze green) on top, white underneath; the female sooty brown where the male is black. Both can acquire darkish grey crowns, napes, and hind necks during summer, sometimes looking as though they've been given whopping great black eyes.

Black-winged stilts don't seem to be very bothered about water pollution and can often be seen (immaculately dressed as they are) wading industriously about in the filthiest and vilest of pools and ponds, unconcernedly picking up worms, molluscs, and insects, occasionally almost up-ending in the water like duck. They are normally found in small flocks (numbers may increase in summer) and fly swiftly, musically calling 'Kip-kip-kip!' to each other as they speed on.

They breed in summer—between April and August—and I have seen their reed-flagged 'nests' (a scrape in the ground) at Sultanpur in years past. Actually, stilts have a history of nesting in the Sultanpur area—apparently huge colonies did so in the salt works that existed in the area over one hundred years ago.

Four cunningly camouflaged, stone-coloured and patterned eggs are laid and the adults make hopelessly neurotic parents. Your approach anywhere near a nest will be greeted by a hysterical dive-bombing attack by both birds, screaming their heads off and trying to lance the top of your head with their bills. The broken wing display is also performed to lure you (and dogs) away from where the eggs are, or from where the chicks are lying doggo.

To my surprise, I discovered that the birds may actually hunt at night too and seem to have no difficulty in locating prey (probably by touch) in pitch-black waters.

Pied Avocet (Avocet), *Recurvirostra avosetta*, Kusiya Chaha

An elegant snow-white and jet-black wader (43 cm) with a slim, upcurved, 'retroussé' bill and long bluish legs. The body is silky white, beautifully set off by the head, hindneck, and two wing bands, which are jet black.

Though avocets do not breed here (only in the Rann of Kutch in Indian limits), they are to be found in the Delhi area nearly all year through. Small flocks may be seen on the Jamuna in the company of other waders (especially black-winged stilts), sweeping their submerged bills from side to side in the shallows, like hockey players dribbling the ball. They pick up small crustaceans, worms, and aquatic insects. Apart from the Jamuna, I have seen them quite regularly at Sultanpur. Whether 'minesweeping' in the water, or flying past, they are an attractive sight, their graceful lines and smart plumage winning them many admirers.

Pheasant-tailed Jacana, *Hydrophasianus chirurgus*, Pihuya

This has always been one bird that makes those sweaty monsoon outings worthwhile. Indeed, the pheasant-tailed jacana (30 cm) is best seen at this time of the year—its breeding season (between June and August), for which it dons special long (15–20 cm), curving black tail feathers. Its plumage is glossy chocolate brown and white, its hind neck gilded oriole gold, its snowy wings black tipped.

Jacanas are also known as lily-trotters and are equipped with enormously long and thin spider toes for lily trotting. The long scimitar tail feathers are dropped outside the breeding season. Like the painted snipe, the jacana is polyandrous, so it is the female that wears the longer tail and her husband who sits on the (usually four) eggs and looks after the brood.

The birds love weed-carpeted stretches of water (and are partial to *singhara*—water chestnut), and may be found on any such patches of water—on the river, or elsewhere around the capital. I have seen them (many years ago, albeit) in the fields adjoining Rajghat, and in the Okhla area. They make for a lovely sight as they fly across the fresh green fields, evocatively calling 'Me-e-oup, Me-oup!' to each other, tails arched gracefully behind.

Bronze-winged Jacana, *Metopidius indicus*, Jal Pipi (Bengali)

Once considered rare enough to be rated as a 'vagrant', the bronze-winged jacana (around 30 cm) can fairly often be seen on and around the Jamuna at Okhla, stomping about on water hyacinth, looking like a waterhen dressed up as a sultan. Its wings are a beautiful bronze green with a dull bullion sheen to them, its head, neck and breast, jet velvet black, its short stub tail, chestnut. A dashing white streak runs from the eye to the nape, set off by a vivid crimson frontal shield and yellow bill. It moves about on enormous spider-toed feet, balancing itself neatly on the wobbly aquatic vegetation. (As with the pheasant-tailed jacana, water chestnut is a great favourite). You really do need a powerful scope to fully appreciate the princely attire of this bird!

Like the pheasant-tailed jacana, it is polyandrous, and breeds between June and September, with the househusbands setting up home on a skimpy pad of reed stems and dutifully raising four chicks. The birds eat insects, molluscs, and the seeds and roots of aquatic plants. The call is a short grunt and a more pleasing, husky piping 'Eek-seek!'.

Little Ringed Plover, *Charadrius dubius*, Zierra/Merwa

A bullet-headed little runabout (15 cm), sandy brown on top, white below with a pigeon-like bill, yellow legs, and a prominent yellow ring around its dark eyes. The forehead is white, the fore-crown, ear coverts, and areas around the eyes, black. A black cravat is worn around the neck, giving way to the white breast and tummy.

These plovers can be spotted spurting about energetically over the muddy banks (or mudflats) of the river or the edges of ponds, tanks and pools, looking like little stop-and-go clockwork toys. They bob their heads rapidly (and typically) as they scuttle about, dead-stopping suddenly to pick up a juicy morsel (insects, small crabs, and suchlike) from the mud. The birds are both residents and migrants to the Delhi area. Residents usually breed between April and June, laying four stone-coloured and patterned eggs on the bare mud, so uncannily camouflaged you could step on them by mistake.

Yellow-wattled Lapwing, *Vanellus malarbaricus*, Zirdi

A tall, upright-standing (26–8 cm) sandy-brown and white bird (of the plover clan) with a white tummy, wearing a black beret and drooping canary-yellow wattles, and looking for all the world like an artist awaiting inspiration. Similar to the more familiar red-wattled lapwing (see following entry) but more elegant and avantgarde.

Not very common in the Delhi area and, so far, I have only seen them (two nesting pairs) within the burly walls of Tughlaqabad Fort. Way back in 1942, three pairs of yellow-wattled lapwings were recorded to be nesting on the Central Vista in the heart of the capital, so the birds have indeed fallen upon hard times since, and have certainly declined.

They are birds of rocky wastelands, and nest in summer—between April and July—laying four (virtually invisible) eggs on the bare ground, and fiercely and noisily dive-bombing anyone who dare approach close. Its call is a rapidly repeated high-pitched (and more neurotic than the red-wattled lapwing's) 'Ti-tee, ti-tee!'.

Red-wattled Lapwing, *Vanellus indicus*, Titeeri/Titiri

The commonest of the plover clan in Delhi and perhaps the biggest nuisance to birders too! The red-wattled lapwing stands alertly tall (33 cm) on reedy yellow legs, and is clad in bark-bronze on top and spanking white below, with a black head, neck and breast. A glistening, dark, blood-like globule—the wattle—gleams in front of each eloquent eye, and a broad white band from behind the eyes curves down the sides of the neck to merge with the underparts.

Red-wattled lapwings can be found nearly everywhere in the city: in the areas adjoining the river, in large parks and gardens, in cultivated areas and wastelands, in woodlands and cemeteries, and even patrolling the expanses of green turf of the Central Vista, the lawns of India International Centre, and the numerous government houses in Lutyens' Delhi. Large gatherings of the clan also take place in the various enclosures of the Delhi zoo. The birds can quite demolish your attempts to get closer to more attractive species at places like Sultanpur, by fluttering over your head and screaming (so famously) 'Did-he-do-it? Did-he-do-it? Did-he-do-it?' as though you were a child-murderer or worse.

These lapwings (like other unprepossessing non-avian species!) appear to have taken a liking to Delhi, and their population has reportedly gone up substantially since the middle of the last century. A scrape in the ground (or a flat terrace) serves as a nest in which four excellently camouflaged stone-coloured eggs are laid. The parents are ballistically neurotic and will attack fearlessly if you get anywhere near their eggs or young, screaming blue murder all the while. But they have their redeeming qualities too. Once, at Buddha Jayanti Park, I spotted a lapwing with six legs. Astonished, I gawped at the bird—standing ramrod straight and eyeing me nervously—and did a re-count. There was no mistake: the bird had six legs! I moved back into cover and peered through the foliage, wondering. Four of the six legs suddenly detached themselves from the other two (and the lapwing) and further split into two pairs, as from under the wings of the bird, two half-grown, fuzzy chicks emerged. As soon as I stepped out into the open, the chicks rushed back, and, hey presto, there was the magic six-legged lapwing all over again!

Lapwings eat insects, grain, and molluscs that they pick up in their typical stiff-backed manner. In Delhi, I have seen them in the lawns at home, Nicholson Cementery, Lodi Gardens, Qudsia Gardens, both the central and northern Ridges, Buddha Jayanti Park, the gardens around Humayun's Tomb, the areas adjoining the river, at Sultanpur, and probably several other places that I have now forgotten Sometimes, at night you can hear lapwings screaming in the dark skies.

White-tailed Lapwing, *Vanellus leucurus*, No local name

A rather plain-looking khaki-brown lapwing (28 cm) with a pigeon face and long yellow legs. Its lovely pleated white tail and lower back, and the slow-flapping broad, white-banded, black-tipped wings are sure pointers to identity. A quiet migrant, arriving by August and staying on till March, this lapwing can be seen at the edges of water-bodies, ponds, and the Jamuna. I have seen it regularly at Sultanpur. It appears to be rather introverted and unobtrusive and rarely draws attention to itself until it flies.

Northern Lapwing (Lapwing or Green Lapwing), *Vanellus vanellus*, No local name

A stocky (30 cm), silent, short-legged bronze-green lapwing with an up-curving crest and dark kohl markings around the eye. This winter visitor (November to February) is not very common, and has been reported from the sandbanks and mudflats of the Jamuna, I have only so far seen them at Sultanpur. On one occasion, a strong, silent flock of about forty birds appeared to be gathering together before their great flight back north, and on another occasion, a smaller, more hungry group of seven birds foraged in a shallow pool. These lapwings fly slowly, beating their broad, rounded wings in a rather erratic manner.

River Lapwing (Spur-winged Plover), *Vanellus duvaucelii*, No local name

A handsome cousin of the red-wattled lapwing, the river lapwing (about 30 cm) is dressed in sandy grey (above) and white (below). It sports a black occipital crest (raised when excited) on its black head, and has a somewhat Napoleonic stance and carriage, with the head and neck tucked down into the shoulders as though into a greatcoat. The tail and rump are white, the former tipped broadly in black. The broad wings are white barred and black tipped, and the legs and bill are black.

Pairs or small parties can usually be seen on the banks of the Jamuna, standing around hunched and somewhat morose, or searching for tidbits in the mud. They breed between April and June on the sandbanks and melon patches alongside the river, and the nest is a shallow scrape in the ground. Three or four beautifully camouflaged eggs are laid. Much quieter than the red-wattled lapwing (though with a similar call), they seem to prefer threatening each other by running stiffly towards one another, wings half-raised warningly.

Small Pratincole (Small Indian Pratincole), *Glareola lactea*, Chhoto Ababeel-batari (Bengali)

A small (17 cm), difficult-to-spot riverside bird, in sandy-brown plumage and with swept-back swallow wings and slightly forked, squarish tail. A delicate black mascara line stretches from the bill to the eye. In flight, a distinct rusty tinge to its underparts (the tummy is white) and the black-tipped white tail are pointers to identity.

Small pratincoles nest (between February and April) in the sandy areas along the riverbank, with colonies of birds squatting down almost invisibly in their shallow scrapes. You can walk right past them blissfully unaware of their presence. Small flocks flicker after insects in flight, especially at dusk, when their white tails flash like lights in the murk. The call is a gecko-like 'Chtuck-chtuck-chtuck!', or a soft 'Tirrit-tirrit-tirrit!' when they are alarmed. Apart from Okhla, I have seen these birds at Sultanpur in summer.

Brown-headed and **Black-headed Gulls**, *Larus brunnicephalus*, *Larus ridibundus*, Dhomra

Both species are winter visitors to the Jamuna and often found together in mixed flocks. The brown-headed gull is the larger of the two (42 cm), dove grey on top, white beneath, with a distinctive white 'mirror' on the two black outermost feathers of its wings. The tail is white. In summer, the head—which is usually white in winter—turns coffee brown.

The black-headed gull (38 cm) is similar except that it has a shining white leading edge to its wings, instead of the white wing mirrors of the former species. In summer it wears a coal-black hood. The bills and feet of both species are red, and they have a curious dolphin-like expression on their faces.

The birds are usually with us from October to April. Recorded as 'uncommon' in years past (the 1960s and 1970s), it looks as though their numbers have increased since, even as the Jamuna gets filthier and more toxic. But perhaps, instead of relying too much on the fish and insects the river still provides, the birds have become increasingly dependent on handouts regularly given by people all winter. Clouds of five hundred or more patrol the river, listening out for the 'Aaao-aaaao!' calls that summon them for a feast. Or else, they simply bob on the water near the several bridges spanning the river, where invariably people stop to feed them.

In fact, it is well worth driving down to one of these bridges (like the pontoon bridge opposite Geeta Colony near Vijay Ghat) to watch the proceedings at dawn. At first, through the pearl-grey winter fog all you can make out (if anything at all!) are the ghostly grey-blue shapes of the birds bobbing on the water. Imperceptibly, the sky tinges to peach and apricot, and the ripples in the water are touched with mauve. Then the sun emerges, frosted like an orange from the deep freeze, setting the wavelets afire. A scooter stops on the juddering bridge, and suddenly you are in a maelstrom of gulls! As one bird they have lifted off the water and winged their way swiftly over, and now swirl like dervishes around your head. The stylish ones snatch the *saeve* and *chidva* scattered at them in great arcs, in mid-air—braking miraculously, snatching, and then flying on. The more conservative birds bob on the lava-gold waters, picking up the manna as it patters down and stabbing out at any other bird within reach. Muted, hoarse 'Ka-yek, ka-yek!' calls fill the air, punctuated by the odd shrill 'Ree-ah!' screams of outrage, as some bandit crow runs a raid through the flock. Then the scooter starts up and departs, and calm returns to the waters, the birds settling down or winging off to explore other possibilities. Until the next benefactor turns up, a few minutes later.

These gulls breed in colonies in the high mountain lakes and bogs of Ladakh and Tibet in summer.

River Tern (Indian River Tern), *Sterna aurantia*, Ganga Cheel

Terns are basically graceful, slim-winged silver-grey and white birds (wearing black velvet caps in summer) that beat swiftly up and down rivers or large water-bodies, looking to pick up small fish.

The river tern (38–46 cm) is one of the commonest of the clan to be seen in Delhi, and patrols lengths of the river either alone or in small flocks. It is a lanky silver-grey bird with a deep papaya-yellow bill, short red legs and feet, and in summer (the breeding season is March to May) has a glossy black forehead, crown and nape. Its shallow tail is deeply forked and has two 'streamer' feathers trailing from them at this time. In winter these disappear and the black head turns white—flecked with black and grey like scattered cigarette ash.

The river tern nests on bare ground—on the sandbanks and mud spits of the river—in colonies with other terns and small pratincoles. Up to three eggs may be laid. The birds have a brisk, no nonsense air about them, with a high-pitched metallic, somewhat laconic-sounding call. They fly purposefully and are always a delight to follow through your binoculars.

Gull-billed Tern, *Gelochelidon nilotica*, Ganga Cheel

Usually smaller than the river tern (38 cm), the gull-billed tern also has a deeply forked tail and is grey and white, but very distinctively has a gull-like bill, which, along with its feet and legs, is black. Indeed, the first time you lock on to one in flight, you could be mistaken into thinking that it is a gull of some kind, or at least a cross between a gull and a tern! In summer, the head, up to the nape of the neck, is jet black. In winter it is white, but streaked with black, and with a black pirate patch over the eye.

This is another fairly common tern of the Jamuna. The birds breed in summer—between the end of April and the end of June.

Black-bellied Tern, *Sterna acuticauda*, Ganga Cheel

Like a smaller, slimmer (33 cm) version of the river tern, the black-bellied tern is best differentiated from it in summer, when its belly, breast, and vent are singed dark grey and charcoal black. Its throat and cheeks are white, the head capped with black, the wings pearl grey, and the tail deeply forked and complete with silver streamers in summer. The slim, long bill is orange-yellow; the legs, orange-red. In winter, the head turns white, streaked with black (there is a black patch behind the eye) and the breast and belly turn white too, flecked with charcoal streaks in some cases.

Thought to be not as common as the river tern, the black-bellied tern is another resident of the Jamuna, and breeds on sandbanks and mud spits in summer. Its call is a shrill 'Kek-kek!'.

Black-shouldered Kite (Black-winged Kite), *Elanus caeruleus*, Kapassi

This small (33 cm), elegant silver-grey hawk (grey above, white below) has black shoulder patches and a black line above the eyes, and burning ruby eyes. When at rest, the wingtips extend beyond the tail, which is white and slightly forked.

It is a bird of open, rocky scrubby country, best spotted on transmission lines from where it likes to keep watch. Else, you may spot it flying intently past and then braking to a standstill in mid-air and hovering, its tail flared out, ruby eyes fixed on the ground, some 10m below. Suddenly it parachutes down to a lower level, hovers again, and then (if it is now low enough) drops on the large insect, lizard, or rodent that has caught its attention.

Never very common in the Delhi area, there are reports that this hawk is getting even scarcer. I have seen the birds over the New Delhi Ridge, in the open country south of Okhla Barrage, and fairly frequently at Sultanpur—especially when the grass is down to harsh tussock stubble. Here, in fact, I once met a family of three kites; the youngster had a brownish wash to its plumage and was screaming vociferously for its breakfast—to be brought to its treetop perch by its parents! Black-shouldered kites breed throughout the year and construct untidy crow-like nests in small trees. Three or four eggs may be laid, and both parents bring up the brood.

Black Kite (Pariah Kite), *Milvus migrans*, Cheel

The familiar dark brown 'cheel' is by far the commonest raptor in Delhi and you can see it circling in the skies from anywhere in the city. More popularly known as the pariah kite, this deft-winged bird (55–68 cm) is a scavenger par excellence that thinks nothing of dodging a spaghetti mess of live wires and cables as well as demented traffic to pick up a run-over rat from the middle of the road. Often ignored by birders because they are so common, these dark brown hawks are superb flyers and worth watching for the air shows they put on—especially in the breeding season (usually winter, but they seem to be extending this to other seasons as well). Pairs or trios take part in high-speed aerial chases, and couples tangle claws and tumble earthward in a somersaulting mass of feathers before disengaging themselves at the last moment and swooping up again. All this to the accompaniment of high-pitched mewling screams.

Kites build large twiggy edifices fairly high up in the trees and often this home is repaired and renovated from season to season, until it attains palatial proportions! Usually two to four eggs are laid, and both parents raise the chicks.

In winter, Delhi's population of resident black kites is augmented by the arrival of migratory kites from beyond the Himalayas. These visitors appear larger than the locals, and their plumage is boldly streaked. They also convey the impression of being the strong and silent types and rarely call out, unlike the rather neurotic-sounding residents. I have counted more than 500 of these birds festooning the trees in Nicholson Cemetery in north Delhi.

An attractive but now rarely seen relative of the black kite is the **Brahminy kite** (*Haliastur indus*, Brahminy cheel). Finished in rusty brown (above), it has an ivory-white head, neck, breast, and abdomen. It keeps to areas near water, so look out for it along the Jamuna. I once got a good long look at one, perched on a dead tree in the enclosure of the African rhino at Delhi zoo.

White-rumped Vulture (White-backed or Bengal Vulture), *Gyps bengalensis*, Gidh

This huge (90 cm) iron-grey and white vulture was once a common sight in Delhi—soaring majestically in the skies or crowding the domes of the tombs in Lodi Gardens. But since the suspected viral epidemic of the late 1990s, these giants of the skies have all but vanished. In overhead flight, the white-rumped vulture could be identified by the broad whitish band under its wings, and while banking or at rest, by its white back or rump. The less common (even in the good old days) **Long-billed Vulture** (*Gyps indicus*) was larger (95 cm) and dull brown rather than iron grey.

I still regularly see (this is at the end of 2000) around four white-rumped vultures circling the skies above Qudsia Gardens and adjoining areas in north Delhi, where I live. In the pre-epidemic days the birds did nest in the huge old trees in the park, and every evening you could see the birds circle and glide down like a squadron of planes returning to base. Vultures are slow breeders, raising only one young every season (usually between October and March) so even if all goes well for the birds from now on, it will take a fair amount of time for these birds to recover their lost numbers and glory.

Egyptian Vulture (Scavenger Vulture), *Neophron percnopterus*, Safed Gidh

A small (60–90 cm), unwashed-looking whitish vulture, with a naked yellow-ochre head and bill, black wing quills, and a distinctively wedge-shaped tail. The birds frequent rubbish dumps and landfills and can often be seen goose-stepping grotesquely on the ground. I have seen them on the mud spits of the Jamuna south of Okhla, as well as in the fields opposite Rajghat—though this was sometime in the early 1980s. The young are dirty brown and can be mistaken for eagles. The birds nest mainly between February and April, stuffing twigs and rags and rubbish into niches in the cornices of old buldings like Purana Quila as I once saw them do. They are known to be partial to offal and human excrement.

Crested Serpent Eagle, *Spilornis cheela*, Furj Baz/ Dogra Cheel

A hefty (88 cm) dark brown eagle with a ruff-like black and white nuchal crest, which it raises like hackles when it is surprised or excited. Best identified in flight by the broad white bands on its dark wings and tail, and the shallow but distinct 'V' in which it holds its wings while soaring. The rusty breast and belly are finely marked and stippled in black and white.

I have seen this magnificent eagle three or four times—all at the same time of the year (September–October and then again in March) and in the same area—the northern Ridge and Nicholson Cemetery. For two consecutive years an eagle arrived at Nicholson Cemetery in the same week in October, and roosted here briefly (not more than a couple of nights) before disappearing. In both cases, it was consistently harassed by crows and black kites. On one occasion, I rescued an injured juvenile, who was being lynched by monkeys—unfortunately the bird did not survive. Perhaps the eagles pass through this area at this time of year while on local migration. The only other place where I have seen this species in the Delhi area is Sultanpur. It is a bird well worth looking out for and it usually perches quite prominently on treetops and exposed branches.

Eurasian Marsh Harrier (Marsh Harrier), *Circus aeruginosus*, Kutar/Kulesir

The marsh harrier (58–68 cm) is a common winter visitor—a raptor that follows waterfowl down from the north and spends the winter harassing these holidaying birds. The sexes are rather differently dressed: the male has silver-grey wings (black edged) and tail, pale rusty head, neck and breast, and is dark rufous below. The female is dark kite-brown all over with caramel-coloured shoulder patches and a matching hood which makes her look rather like a bandit about to pull off a heist. Plumage may vary confusingly depending on the age of the birds. The harriers are somewhat stockily built and fly with icy deliberation, beating their wings purposefully and then gliding from time to time, wings held in a shallow 'V' above the body. They patrol waterside habitats where waterfowl congregate, often causing panic amongst resting birds and literally harrying them from one side of the water body to the other. The idea is to spot a laggard or weakling in the flock and then drop down on it. The birds can often be seen on the ground too, keeping watch from a tussock or hump, for frogs, small reptiles and animals, and large insects, which also form part of their diet.

Sultanpur was a great place to watch marsh harriers at work. They would arrive by August and stay on till April, patrolling the jheel and its surrounding woodlands more assiduously than any forest guard. The very sound of hundreds of panicky wings flurry-splooshing in the water (like some giant rinsing out his mouth) was a sure sign that the harrier was out hunting. On several occasions I have come across the tattered remains of a kill. Harriers can also be seen along the Jamuna—and, again, if you ever hear a whole flock of ducks trying to take off at the same time, look sharp. Somewhere in their midst the march harrier flies, its cold eyes searching for some hapless, silly coot.

Shikra, *Accipiter badius*, Shikra (male)/Cheepak (female)

This fierce-eyed, diminutive (30–6 cm) woodland hawk is

common in Delhi's parks and gardens and green areas. Males are smaller than females, clad in bluish-grey suits (above) with a fine, horizontal rusty weave on their breasts. The females are more brownish-grey above but otherwise quite similar. This bullet-headed little hawk has a small but fiercely hooked bill and sunflower-yellow (or burning-orange) rings in its eyes. Youngsters have dark vertical streaks down their creamish breasts.

The shikra waits in ambush in wooded cover and dives swiftly after its target, beating its wings rapidly to gain speed, then gliding with great verve as it twists and turns dexterously between the branches and trees in pursuit of some hapless dove. Sometimes it is worth waiting at feeding spots in parks, which attract its prey—squirrels, mice, and small birds—especially if you know that a shikra is in the vicinity. You may be treated to a live demonstration of its hunting skills. The call is a shrill 'Ke-kee! Ke-kee!' which usually gets sparrows, mynas, and doves all panicky and excited. Shikras nest between February and June, building an untidy platform of twigs (lined suitably) high up in some tree. Three or four eggs may be laid and both parents bring up the brood.

Oriental Honey Buzzard (Honey Buzzard), *Pernis ptilorhyncus*, Madkare

About the size of a black kite (60 cm), the oriental honey buzzard is a greyish-brown raptor with a dark grey head and pale brown underparts, cross-barred with white. Its plumage is however very variable, so look for a short black crest and scaly-looking feathering about the head (which tends to get ruffled by the breeze). The wings are silver grey, closely barred on their undersides. Often it is difficult to make out the exact colours, as the bird tends to sit somewhat lumpily in the dark shade of woodlands and groves. In flight, the wings and tail appear broad and long, the head curiously small and cuckoo-like. The wing beats are steady and regular, the flight interspersed with brief glides when the wings are held horizontal with the body.

The honey buzzard is a woodland tree-loving bird that spends long hours brooding in the gloom of the canopy. I have seen it in Deer Park (near Hauz Khas)—appropriately enough near a beehive—and, on several occasions, on the northern Ridge. An inveterate honey stealer, the bird raids the hives of honeybees, happily impervious to stings, thanks to the scale-like feathers around its head. Its call is a high-pitched screaming whistle. Honey buzzards nest between April and June, and two young are usually raised by both parents.

Long-legged Buzzard, *Buteo rufinus*, Chuhamar

Also about as large as the black kite, the long-legged buzzard is overall usually of a much lighter shade than the honey buzzard. Again, its plumage is very variable, in shades of brown tan and dun, its head, neck, and breast usually paler, even creamy white or caramel. Its tail is pale rufous; the legs are unfeathered.

The bird is a winter visitor to the Delhi area, though not very common. I have seen it a few times on the northern Ridge; on one occasion, back in the early to mid 1980s I got a good look at one as it sat on a dead tree trunk and seemed to meditate. Its plumage was beautifully stippled and marked, its head creamy gold, the beak, small, neat, and not very dangerous looking. All too soon it was discovered by crows that began harassing it, and it eventually flew off with lumbering wing beats, through the canopy.

The countryside around Okhla and the riverine area would be other good places to look out for these buzzards—they like perching on mounds or posts as they keep watch for lizards and small rodents.

Tawny Eagle, *Aquila rapax*, Ukaab

A hefty (63–70 cm) umber to buff-brown eagle equipped with trademark grappling-iron talons and hooked bill. The head is usually a shade paler than the rest of the plumage. The tail is rounded.

This resident eagle has been described as being fairly common in the Delhi area in the past, and was found nesting is such clamorous localities as those near the railway bridge off the Red Fort. I have, so far, only seen it on mudspits and trash dumps in the Okhla area, and that too way back in the mid 1980s.

Common Kestrel (Kestrel), *Falco tinnunculus*, Karontia

This lovely little falcon (32–5 cm) has an ash-grey head and brick-red body, attractively patterned with dark brown flintheads. The underside is pale caramel or buff, and the head of the female is pale sandy and not grey. The tail is grey with a black band near the tip.

Kestrels are winter visitors and can be found flying swiftly over fallow fields and open country, using shallow wing beats interspersed with short glides. Their wings are long and slim. Often they will fly quite low over fields and sometimes treat you to a spectacular display of hovering—rather in the manner of the black-shouldered kite. A bird will park itself 10 to 15 m up in the sky, intently scanning the ground below, its wings beating rapidly, tail fanned out. A lizard, field mouse, grasshopper, or any such delicacy, if spotted, stands no chance. The kestrel pounces straight down and wings swiftly away with its victim.

Probably the best places to spot these birds are the open areas south of Okhla Barrage. For one, there is a vast skyline for you to scan here, and, second, the fields and open country are prime hunting grounds for this attractive little raptor. I have also seen these birds at Sultanpur, when the grass was dry and cropped short.

Little Grebe (or Dabchick), *Tachybaptus ruficollis*, Pandubi

A rotund, fatty little water-bird (25–30 cm) with no tail, all puffed up rear end, and staring golden eyes, the little grebe is an endearing little creature indeed. Its underparts are silky cream white, and in the breeding season (April to October) the head and neck are a lovely shade of dark maroon or wine red. The fleshy gape is bright yellow and there is a prominent white spot on the bill.

Little grebes can be usually seen on the Jamuna, puttering solemnly about in the company of other water-birds (and their own kind), and diving beneath suddenly, leaving only expanding hoops of water behind. I have seen them regularly at Sultanpur too, where they breed, at the zoo and in Badkhal Lake. Any self-respecting water body around the capital should play host to these little birds. Their calls, which sound like a shrill, trilling alarm clock (or football referee's whistle) going off, are a sure indication of their presence. Look keenly at the reed-fringed banks and edges of the river or water body and you will be rewarded. The nest is a soggy wad of reeds on floating vegetation often half submerged. Three to five eggs are laid and often you can see the chicks paddle strenuously after their parents. When tired or frightened they simply clamber on to their parents' backs and hitch a piggyback lift, often vanishing right into the plumage.

Little grebes eat aquatic insects and larvae, tadpoles, frogs, and such like fare. They may entertain themselvs by chasing each other over the water, beating their stubby little wings splashily, half running and half swimming, and, evidently, enjoying themselves.

Darter (Indian Darter), *Anhinga melanogaster*, Panwa

A deadly-looking, snake-necked water-bird (90 cm) in shimmering black with silver-marked wings and back, a velvety brown head and neck, and off-white chin and throat, the darter is one of the more elegant of the fish-spearing clan. Its bill is sharp and javelin-like, the tail long and stiff and wedge-shaped.

Darters are usually found in the company of cormorants (see following entry) though in the Delhi area they are nowhere nearly as common. A few may be occasionally spotted in the water-bodies and acacias in Delhi zoo, especially just before and during the monsoons, when the birds breed. You may also see them flying over the Jamuna in the Okhla area, or at Sultanpur and Badkhal.

Darters are also called snakebirds because of their ability to swim semi-submerged, with only their serpent-like necks cleaving the water's surface. Inveterate fishermen, they capture their prey by diving under and giving chase, and scissoring rather than stabbing at their prey with their long spear-like bills. Their plumage, like that of the cormorants, is not waterproof, so wings must be hung open to dry in between fishing expeditions—or else, being soggy and heavy, can pull the bird under water and drown it.

Cormorants

These familiar jet-black water-birds can be regularly seen flying over the river or other water-bodies around the capital, as well as swimming low in the water together as they launch their cooperative fishing expeditions. Three kinds of cormorants can be seen in the Delhi area.

The **Little Cormorant** (*Phalacrocorax niger*), called the jalkawwa (water crow) as all cormorants are, is about the size of a large duck (50 cm), and can best be met at the Delhi zoo, where the birds breed during the monsoons in busy colonies. Silky black all over, the bird has a somewhat duck-like profile, but a sharply hooked, compressed-looking bill makes certain it is no duck. The tail is stiff and long.

Little cormorants are fairly common at most water-bodies around Delhi and I have even, quite regularly, met a rather sad and morose-looking pair in the tiny ponds of the northern Ridge. The Jamuna, Badkhal, and Sultanpur are other places where you can meet them in much greater numbers.

A little larger in size (at 63 cm) is the **Indian Cormorant** or Shag (*Phalacrocorax fuscicollis*), which also has a larger and slimmer bill than the little cormorant, and blue-green eyes. Again, the Delhi zoo is the best place to make your acquaintance with these birds and compare them to one another.

The **Great** or **Large Cormorant** (*Phalacrocorax carbo*), at between 80 and 100 cm, is indeed the big boy of the trio. Its throat and the front half of its face are white. Bulkier than the other two, it has a thick neck and short tail and flies with steady (as against somewhat fluttery) wings beats. Like the other two, it is jet black all over, though in the breeding season (during the monsoons), it sports two oval white patches on its thighs. Large cormorants may be seen in the riverine area, passing time on the acacias gazing vacantly into space, their gular patches fluttering as they attempt to keep cool. I have regularly seen them flying over the Badkhal area and at Sultanpur.

All these cormorants feed chiefly on fish, frogs, and crustaceans, diving beneath and swimming swiftly underwater as they give chase. They are gregarious birds and all three species (and darters as well) can often be found nesting and even fishing communally. Like the darter, cormorants also must dry their wings and plumage in between fishing trips to prevent waterlogging!

If you live in, or pass by the areas around the Jamuna, you cannot miss seeing these witch-black birds flying above in disciplined 'V' formations or in echelons strung across the sky, every morning and evening as they make their way to and from their fishing and roosting areas.

Egrets

Tall, dagger-billed, angel-winged, and satan-eyed, egrets are those familiar white birds that can be seen stalking along river banks, or in fields, or even on garbage heaps, in and around Delhi. There are four kinds common in the Delhi area, and we will take them in descending order of height.

The tallest, at between 65 and 72 cm, is the **Great** or **Large Egret** (*Casmerodius alba*, Bara Bagla). Pure white, it has a yellow bill and during the breeding season (the monsoons) has lacy plumes drooping waterfall-like beyond its tail. At this time, the bill turns black and the bird sports a bright blue-green eye patch. Not as

Plate 32   *Pond Heron*   In fine breeding regalia.

Plate 33   *Purple Heron*
The riverside is the best place to appreciate flypasts.

Plate 34   *Little Heron*
Uncommon crepuscular skulker.

Plate 35 *Black-necked Stork* Getting rare!

Plate 36 *Bay-backed Shrike*

Plate 37 *Grey Shrike*

Smart butcher birds all!

Plate 38 *Rufous Tree-pie* May I ask you for 'thocolate'?

Plate 39 *Small Minivet*   Seen flitting about like glowing embers in the canopy of parks and large gardens.

Plate 40 *Long-tailed Minivet*   Stunning, but none too common.

Plate 41 *Black Drongo* Gives protection to gentler species, but not averse to taking *hafta*.

Asian Paradise Flycatcher (Male & Female)
Probably a summer and monsoon visitor to the capital, especially around water-side habitats.

Plate 42 ♂

Plate 43 ♀

Plate 44 *Black Redstart* Genteel winter guest in parks and gardens.

Plate 45 *Common Stonechat*   Winter visitor to open country.

Plate 46 *Brahminy Starling*   Spirited, rakish, resident songster.

Plate 47 *Asian Pied Starling*   Resident in gardens and open areas.

Plate 48 *Wire-tailed Swallow*   Best seen during the monsoon, near water.

Plate 49 *Red-whiskered Bulbul*   The cheerful songster!

Plate 50 *Red-vented Bulbul*   Familiar, pugnacious resident of every garden and park.

Plate 51 *Ashy Prinia* Skulks about the bottom of hedges.

Plate 52 *Jungle Babbler*   Always turning up the dirt; cross-looking but endearing.

Plate 53   *White-browed Wagtail*   Usually found near water bodies, and especially on boats!

Plate 54 Red Avadavat or *Red Munia*   Resplendent in the monsoon.

Plate 55 Silverbill or *White-throated Munia*   Found in fields, on the Ridge, and in gardens (if given all meals!)

common as the other egrets, this tall and dignified bird can be spotted along the river bank, usually intent on its fishing.

The slightly shorter **Intermediate Egret**—also known as the **Smaller** or **Median Egret** (*Mesophoyx intermedia*, Korch Bagla)—is difficult to distinguish from the great egret since there is a lot of variation in the height of both these species. It does have a shorter bill though, and in the breeding season wears lace on its breast as well as its back. The yellow bill now turns pure black and the bird has no crest.

The **Little Egret** (*Egretta garzetta*, Kanchia Bagla), at between 55 and 65 cm, is slim and graceful and has prominent black legs with bright yellow feet, which help to distinguish it from the otherwise similar intermediate egret. In the breeding season it sports two lacy white plumes on the top of its head as well as a fountain of lace cascading down its back well beyond the tail. Its sharp, pointed bill is black.

The **Cattle Egret** (*Bubulcus ibis*, Surkhia Bagla) at between 48 and 53 cm, is the smallest and stockiest of the lot and can often be seen well away from water. Its yellow bill is relatively short and less dangerous looking than those of the other egrets. During the breeding season it is easily identifiable by its lovely saffron head, neck, and back.

The Delhi zoo is a good place to become familiar with egrets, especially the intermediate and little egrets that breed in colonies in the acacia clumps on the various island enclosures. You can spend hours watching them court, nest-build, and try to bring up their dreadfully bratty broods. Cattle egrets can often be met in fields, usually in the company of livestock, whose footsteps they dog in order to snap up the insects disturbed by the beasts as they graze.

Egrets can also often be seen flying across the capital's skies, especially at dawn or dusk, and they make a lovely sight as they float ethereally past, wings translucent in the bright sunlight, heads racked back into their shoulders in easy rider fashion. The birds are usually silent, except when upset or disturbed, when they 'Quaark!' hoarsely with irritation. They feed on fish, frogs, and insects, expertly caught with patience and javelin-like thrusts of their bills.

Indian Pond Heron, *Ardeola grayii*, Bagla

This stocky, relatively short-legged heron (42 to 45 cm) is quite common around water-bodies and usually quite invisible too—until it takes off and flies. Earthy brown, with dried-grass-like streaking on its neck and head, the heron stands stock-still at the edges of water-bodies, until it decides that you have come too close. Then, with an irritable 'kwaark!' it unfurls dazzlingly white wings and makes a low and slow exit over the water. During the breeding season—the monsoons—the pond heron undergoes a stunning transformation and now decks up in maroon lace (cascading down its back) and sports two white, filament-like plumes on the back of its head. The black-tipped bill may be blue grey, yellow or orange, and the scaly legs yellow or reddish.

Pond herons, or paddy birds, may be easily seen on the banks and mudspits of the Jamuna, at the zoo, Sultanpur, Badkhal, and even at the watercourses at Lodi Gardens and Buddha Jayanti Park, not to mention those running alongside Rajpath. I have also seen them—especially during the monsoons—at the ponds on the northern Ridge. They look for fish, frogs, crabs and insects, and nest in colonies (at the zoo for instance) along with other herons like the black-crowned night herons.

Purple Heron, *Ardea purpurea*, Lal Anjan

A tall (78 to 90 cm), lanky thundercloud-purple heron with a sinuous chestnut neck and stunning resin and black undersides. Somewhat secretive, this heron conceals itself well amongst the cattails and rushes of the riverside and is quite difficult to spot. Usually you will flush it by mistake whereby it will spring into the air and flap ponderously away. These herons can be seen in the swampy, cattail-flagged banks of the Jamuna, near Okhla, as well as at Sultanpur and Badkhal. They breed in small exclusive colonies during the monsoons, and the young—three to five per clutch—are uniformly cinnamon brown.

Grey Heron, *Ardea cinerea*, Nari, Anjan

Larger than the purple heron (100 cm), the grey heron is dressed in a formal silver-grey and white suit, has a yellow dagger bill and sports a black cut-out skullcap and occipital crest. The crown is white, the legs and bill are yellow.

Quiet and serious fishermen, grey herons can be spotted lining up on the riverbanks in the Okhla area (and probably elsewhere too)—leaving decent spaces between one another. They are more active in the evening. They fly with steady wing beats, heads drawn back into their shoulders, legs trailing behind.

Grey herons breed during the monsoons, when three to six eggs are laid in a twiggy platform nest.

Little Heron (Little Green Heron), *Butorides striatus*, Kancha Bagla

This stocky, hunched, somewhat morose-looking heron (40–8 cm) is something of a rarity in Delhi, and though it has been reported from Okhla, so far I have only come across a single pair at a pond on the northern Ridge, opposite Hindu Rao Hospital.

This heron's greenish-black feathers are beautifully outlined with ivory, and its black crown is adorned with a long glossy black crest. The pair I discovered on the Ridge was characteristically secretive and crepuscular and must have been nesting somewhere in the thick tangle of acacias crowding the pond. When startled by my dog they flew across the water with an irritable 'K-yow! Ke-yek!' and vanished in the gloom. Little herons breed during the monsoons, building a shallow nest of small twigs a few metres above the water. Both parents look after the brood.

Black-crowned Night Heron (Night Heron), *Nycticorax nycticorax*, Kokrai/Kwak

This dusk-loving heron (58–65 cm) can best be seen on the acacia-shrouded 'islands' at the Delhi zoo between June and September, when colonies of the birds set up home alongside the nesting egrets and cormorants. Smartly clad in glossy blue-black above, white below and with ash-grey wings, this heron usually crouches deep in the depths of foliage, staring unblinkingly at the world out of ruby-red eyes. A fashionable crest of thin white plumes droops behind its head during breeding season. In flight, its broad round wings, and squat bull-nosed silhouette make it unmistakable, as does its hoarse 'Waark!' croak.

Black-crowned night herons are most active around dusk, though those at Delhi zoo are quite happy to feed actively alongside painted storks, egrets, and even adjutant storks in brilliant mid-day, if the fish is good. The birds also nest in reed beds and can often be seen flying over the Jamuna. I have also seen them at Sultanpur and Badkhal Lake, though really the zoo remains the best bet. While the male does the fetching and carrying of nesting material, the female puts together the nest and both parents bring up the young. The juveniles are brown, with sharp off-white markings and bright golden eyes.

Bitterns

Shy and secretive to an extraordinary degree, bitterns are best spotted as they fly from one hiding place in the reed beds along the banks of the Jamuna, to another. July and August are the best months for bittern spotting: look out for stocky, heron-like birds flying in bit of a hurry with flurrying wing beats. If you can track them down to their hiding places in the reeds, you may see them adopt their typical 'soldier-on-guard' postures; stiff and ramrod straight, and willing you to believe that they are not there.

The **Yellow Bittern** (*Ixobrychus sinensis*, Jan Bagla) is a small (38 cm) slim-necked, narrow-winged, yellowish-fawn heron, with pinkish tinges to its plumage and black flight feathers. Its legs tend to dangle clumsily while in flight.

The **Cinnamon Bittern** (nee Chestnut Bittern, *Ixobrychus cinnamomeus*, Lal Bagla), about the same size as the yellow bittern, is startlingly attractive in chestnut rufous. Its throat is creamy with a dark stripe down the foreneck and streaks on its lower breast and abdomen. Females are duller and have a blackish crown and more prominent streaking on the breast and throat. The birds breed during the monsoons, the yellow bittern on a pad of reed flags near water, and the cinnamon bittern on a small twig platform in reeds adjoining a swamp or in bushes next to small water-bodies. Both parents look after the four to six chicks.

Greater Flamingo, *Phoenicopterus ruber*, Bog Hans/Rajhans

I will never forget the distraction caused to me when, while driving down Ring Road near Rajghat, I suddenly spotted a trio of flamingos through the windscreen, as they flew languidly high over the car, overtaking it with disdainful ease. But it was nice to know that these graceful, comical ballerina-like birds were probably 'putting up' somewhere on the river near where I lived. More recently a flock of flamingos, numbering between 100 and perhaps 250 have been camping for several months in the shallows and mud banks of the Jamuna at Okhla, and have been drawing admiring groups of birders.

They are greater flamingos, standing tall and proud between 125 and 145 cm, on slim, long, ballet-dancer legs (but with ungainly webbed feet), dressed in short white tutus flushed with pale pink. Their wings, when unfurled, are deep pink and black, their comical banana-shaped bills (which are highly specialized for filter feeding in the upside down position!) are pale pink.

In the past flamingos have been reported from the jheels around Delhi but not so from the river itself. In the mid 1980s, a flock of over 500 birds stayed on at Sultanpur for several months. A single **Lesser Flamingo** (*Phoenicopterus minor*), shorter, stockier, and a darker pink, minced stiffly along with the flock. Flamingos thrive in brackish and alkaline shallow-water-bodies, where they find molluscs, crustaceans, larvae, and algae to their liking. They breed in the vast wastes of the Rann of Kutch.

Ibises

All three species of ibises found in India can be seen in the Delhi area. The **Black-headed Ibis** (nee White Ibis, *Threskiornis melanocephalus*, Munda) is easily the most common of the lot. A large (75 cm), hunched, stork-like white bird, with a stout down-curving black bill and black legs, and leathery-looking face, this ibis can best be seen at Delhi zoo, where, during the monsoon, it breeds alongside the colonies of egrets, night herons, and cormorants. A startling scarlet streak under its wings (in the armpits!) is prominently visible at this time, when the bird flies. Black-headed ibises can also be seen at other water-bodies such as the river, Sultanpur, and Badkhal.

The smaller, stockier **Black Ibis** (*Pseudibis papillosa*, Kala Baza), standing hunched at 68 cm is less frequently seen. I have, in fact, only occasionally seen them at Sultanpur. Dark brownish-black in appearance, they have prominent white epaulettes (especially visible while flying), a rust-red skullcap (actually a crop of warts!), and reddish legs. They fly in a strange slow-fluttery manner, rather like fruit bats.

The **Glossy Ibis** (*Plegadis falcinellus*, Chhota Baza) is the smallest (55–65 cm) and most graceful of the trio. Its down-curved bill is slender, its dark mother-of-pearl plumage shot through with metallic purple and green. The head and neck are deep chestnut, and, especially from a distance, the birds look broodingly dark. Small flocks of glossy ibises can sometimes be seen during the monsoons in flooded fields and ditches alongside roads on the outskirts of Delhi. I have seen them at Sultanpur too.

Ibises eat insects, grain, small reptiles, molluscs, and crustaceans, picking them up deftly with their tweezer-like bills.

Eurasian Spoonbill (Spoonbill), *Platalea leucorodia*, Chamcha Baza

A large (80–90 cm) but graceful snow-white marsh bird with a long, distinctive, yellow-tipped black bill somewhat reminiscent of a soup ladle. During the breeding season (between July and November), the birds wear a bushy, ponytail-like white crest and stains of sulphur on their breast. However, it is not known whether spoonbills breed in the Delhi area.

They can, however, be seen in small or large flocks on the Jamuna or at water-bodies like Sultanpur, often in the company of other waders. They feed actively in the mornings and evenings, advancing through the water in a broad front, necks craned forward, bills scything through the murky suspension in a semi-circular motion. Sometimes they appear to be guided and driven by compelling underwater forces as they slew side to side in an erratic zig-zag manner, making their way purely by feel as they pursue frogs, tadpoles, and insects. (they also eat vegetable matter). Spoonbills fly with graceful dignity—often in great 'V's festooned across the sky—necks stretched out, wings beating steadily, translucent in the sun.

Great White Pelican (Rosy Pelican), *Pelecanus onocrotalus*, Hawasil

The Delhi zoo is the best place to get familiar with these big, wacky birds with goofy-grins. A flock floats around the lake like a flotilla of flying boats, their dazzling white plumage glowing pale gin-pink, their huge shopping-bag bills cleaving the water or flapping like sails to keep cool, their small button-eyes puckish, and their crests utterly absurd. While some of these birds are pinioned, others can fly and they will, at office hours, line up and take off splashily one after the other, displaying their lovely black flight feathers as they spread their huge, broad wings. Gaining height on the thermals, they soar in the heavens and then plane down, neatly furling their wings by their sides like rolled-up umbrellas.

It is also worth watching them fish as a team. Their pincer movement harries the hapless fish into the shallow where they dip those big bag-like bills of theirs like nets and scoop up the catch. (Their bills indeed can hold more than their bellies can!)

Pelicans can also be seen at Sultanpur, usually in winter, especially if word has got around that the fishing is good. It is wonderful watching a flock of a hundred odd birds drift silently overhead, back and circle casually, and then whoosh down to land, bobbing cork light on the waters.

Painted Stork, *Mycteria leucocephala,* Janghil/Dokh

This tall, decorative, yet slightly oafish-looking stork must surely be one of Delhi's favourite avian inhabitants, especially amongst visitors to the Delhi zoo. Standing tall (though usually stooping!) at around 1 m, the painted stork has a huge waxy yellow swordbill slightly downcurved at the tip, a domed head the colour of an old cricket ball, and white plumage beautifully filigreed and dappled in greeny-black lace. A flush of candyfloss pink adorns the shoulders and wings, and flight feathers are metallic greeny-black. The long bare legs are pinkish.

The acacia-clad islands at the Delhi zoo have for long been used for nesting by colonies of painted storks who arrive in around August and stay on till March. This provides a great opportunity to observe the family life of these birds from close quarters.

Nest-building commences immediately on arrival; huge platform-like nests are constructed (or old ones renovated) amidst much arduous bill-clattering, and soon three to five eggs are laid—much to the mouth-watering delight of the gangs of ruffian crows hanging around. Eventually the chicks hatch; coal black and cotton-wool white and dreadfully lugubrious. As they grow, they look increasingly unprepossessing, their plumage turning dirty grey, the demeanour almost ghoulish. Like all fledglings they have colossal appetites and soon are experimentally flapping their dirty grey-black wings while standing at their nests and imagining great conquests of the skies! Eventually they will join their parents in flight. Adults can often be seen soaring in wide circles high above the capital, rather in the manner of vultures.

Painted storks eat fish, frogs, and insects found in water, those at the zoo often picking up fairly large fish. The birds can also be seen on the Jamuna, and of late, they have attempted nesting (with modest success) at Sultanpur.

Asian Openbill (Open-billed Stork), *Anastomus oscitans*, Gungla

This stork is not too common in the Delhi area and I have only seen it on the Jamuna near Okhla and at Sultanpur. A stocky, 68-cm-tall stork, in smoky greyish-white plumage, with black wings and tail, the bird's claim to fame is its strange eye-of-the-needle-like bill, in which the two mandibles do not sit flush on each other. The gap is thought to help the bird is some way in separating the soft juicy bodies of snails (which is what they thrive on, apart from frogs and crabs and other such creatures) from their shells. During the monsoons, when they breed, the smoke-grey plumage turns pure white.

Woolly-necked Stork (White-necked Stork), *Ciconia episcopus*, Laglag

Though recorded as a common resident in the 1970s, I have seen the woolly-necked stork only infrequently at Sultanpur, either alone or in small groups. With its black body glossed with purple-green and copper, 'woolly' white neck and black skullcap, and straight blackish bill, this 85-cm-tall stork is unmistakable. Given to nesting very high up on trees—silk cotton or semal, for example—during the monsoon, this stork hunts fish, frogs, and such like fare. Three to four eggs are laid and both parents share all the duties.

Black-necked Stork, *Ephippiorhynchus asiaticus*, Loharjung

This almost 1.5-m-tall implacable black and white stork, with its reedy coral-red legs and massive black sword bill was not so long ago—in the 1960s and 1970s—seen quite commonly along the river banks and jheels around Delhi. As many as forty birds were reported from one gathering of the clan at Najafgarh. Today you will be fortunate to spot a pair of these taciturn storks at Sultanpur perhaps, and, if you are really lucky, a juvenile along with its parents.

At close quarters the bird's black neck and feathers shimmer with metallic tints of inky blue, purple, and green. Females have hard yellow eyes, males soft brown ones. In overhead flight, a broad black band through the pearl-white wings is prominent; the tail is black.

Black-necked storks nest between August and January, constructing huge twiggy edifices on the tops of large trees, usually near water. Three to four white eggs are laid. Though I have seen juveniles with their parents regularly at Sultanpur, I've never found out where their home was. These storks eat fish, crabs, frogs, and, often, other water-birds such as coots.

You can also see black-necked storks at Delhi zoo, though these are pinioned birds, unable to fly.

Greater Adjutant (Adjutant), *Leptoptilos dubius*, Garur

This massive, 1.5 m choleric iron-grey stork, with its naked flesh-pink face, colossal yellow wedge-like bill, and disgusting sausage-like appendage dangling from its chest, was seen in fairly large numbers (over 100 birds at a time) in the water-bodies around Delhi in the 1950s and 1960s. I have, however, only met these crusty old ladies and gentlemen at Delhi zoo, where they have had their flight feathers removed and so can only amble around on foot.

Apart from consuming fish and crabs and small animals, the birds are excellent scavengers and so, it is said, during the bitter rebellion of 1857 cleaned up a lot of the dead bodies that lay strewn around the Delhi Ridge.

Shrikes

Smart, deadly, and masked, shrikes are soft-plumaged, large-headed, hook-billed birds that can be seen keeping vigil from exposed perches, bush-or treetops, transmission lines, or anywhere else suitable, for large insects, lizards and frogs. Loners, except during the breeding season—which lasts from February to October—they are harsh-voiced birds who only mellow while courting—when they produce pleasant warbling melodies and show off their expertize as mimics. The nest is a cobweb-bound deep cup made of spiky twigs, suitably lined with feathers and rags into which two to six white eggs, splotched with purple are deposited. The female incubates while both parents feed the brood.

The **Bay-backed Shrike** (*Lanius vittatus*, Pachanak) is probably the most avant-garde of Delhi's shrikes even if the smallest is at just 17 cm. A silver-grey head contrasts beautifully with a deep maroon-burgundy back. The wings are jet black with white mirrors, the tail black and white, the throat white, faintly washed in buff, and the belly pale buff. The bay-backed shrike can regularly be seen in parks and open country in and around Delhi, such as the New Delhi Ridge, Tughlakabad, and Asola Sanctuary, for example.

The **Long-tailed Shrike**, earlier the Rufous-backed Shrike (*Lanius schach*, Maltiya Latora) at 25.5 cm is obviously larger and has an autumn-rufous lower back and rump. This shrike, perhaps not as common as the others, can be seen in thorny, prickly country, usually near water, for example around the reed beds of the Jamuna and at Sultanpur.

The **Great Grey Shrike** (*Lanuius excubitor*, Dudhiya Latora), the same size as the long-tailed shrikes, is silver grey and black and white, and usually likes to keep watch over stony dry areas and scrub country from transmission wires. I have often seen it in the open country adjoining Okhla, at Sultanpur, and at Badkhal.

Shrikes often hunt more than they immediately need to eat; surplus kills are impaled on thorns and saved for later, and for this reason they are called butcherbirds.

Rufous Tree Pie (Indian Tree Pie), *Dendrocitta vagabunda*, Mah-lat

The tree pie's flute-clear 'Ko-ki-la!' or 'Bo-bo-link!' call can usually be heard in parks and gardens all over Delhi (often followed by a harsh, metallic, grating squawk) much before the bird can be sighted. An arboreal resident, the tree pie is handsomely turned out, with a sooty head and throat, cocoa-rufous back, pale grey and white wings, and long, glittering silver-grey tail with a broad black tip.

Often pairs or small family parties will flit musically through the canopy, feeding on fruit, berries, insects, small animals, and even baby birds. Sometimes they come down to feed—especially where bread and other fare has been scattered for birds. Buddha Jayanti Park is one place where you can almost always see tree pies. Other favourite haunts would include Lodi Gardens and Deer Park, both the central and northern Ridges, and really any park or garden where there are plenty of trees. Tree pies nest mainly between March and May; four or five white eggs, streaked and splotched with bright reddish brown are laid in an untidy but strong crow-like nest.

House Crow, *Corvus splendens*, Kawwa

This big (40 cm) inky-black rascal, with its glittering eye and iron spike bill really needs no introduction. The numbers of crows in Delhi appear to have increased enormously judging by the large, noisy, cleaning-up squads one sees in places like Lodi Gardens after a picnic party has departed. While crows are ruthless serial killers (and consumers) of baby birds everywhere, their contribution to keeping the city's streets and lanes clean cannot be underestimated. Also, they have character and personality in spades, and that special street-smart savoir-faire that makes them do so well living with us. Crows breed between May and July—and, for all their smartness, are often cuckolded by koels.

Large-billed Crow (Jungle Crow), *Corvus macrorhynchos*, Karrial

Larger (46–59 cm), glossier, hoarser, and more dangerously armed than the house crow, the all-black large-billed crow is much less common than the former. Its status and numbers in Delhi, however, seem to have improved considerably since 1950 when it was listed as 'vagrant' and seen in small numbers only sporadically. Today, any large rabble of house crows is almost sure to have a few pairs of large-billed crows in its midst. These 'birds in black' are quieter and more evasive than their grey-collared cousins, though ruthlessly deadly with the young of other species. I have seen these birds regularly on the northern Ridge, and at Buddha Jayanti Park, Lodi Gardens and Delhi zoo.

Eurasian Golden Oriole (Golden Oriole), *Oriolus oriolus*, Peelak

Exquisite in laburnum gold and black, the golden oriole (25 cm) with its sleek effete lines and brightly lipsticked bill is one of the star birds of the Delhi summer. It is thought to arrive in the capital around early March and will stay on till October by which time it will have courted, nested, and seen its fledglings safely off.

The male is bright golden yellow, with *kaajal*-like black eye make-up and a vivid pinkish-red bill—and really quite effeminate looking. The female is a duller olive yellow. Orioles are shy arboreal birds, often difficult to see in spite of their dazzling colours. They give their presence away by their haunting musical 'Peeela-wee-o!' song. The female builds a small, snug, hammock-like nest into which she just about fits (while her husband stands guard and sings) and incubates three to four black-spotted white eggs. The birds eat insects, berries (especially figs), and nectar. In Delhi, golden orioles may be seen in tree-shrouded gardens and parks everywhere—though not in large numbers.

Small Minivet, *Pericrocotus cinnamomeus*, Balalchashm

A high-pitched 'Tswee-twsee-tswee!' call at the very edge of your upper hearing scale is a sure indication of the presence of small minivets. In their ember colours of soot black and ash grey, lava red and flame yellow, they flicker in small restless groups amidst the acacia canopy like tiny restless flames. Females lack the charcoal-grey and sooty-black colours of the male, and are yellow whereas the males are red.

Happily, small minivets (15 cm) are residents of Delhi and can be found on both the Ridges, in parks such as Lodi Gardens, Buddha Jayanti Park and Deer Park, and even large private gardens. They are restless insect hunters, not as difficult to spot as to keep track of as they flit in small parties through the canopy.

Small minivets build tiny, cup-shaped nests high in the trees, bound together with cobweb. Two to four pale sea-green eggs, splotched with reddish brown or lilac, are laid and both parents look after the brood.

Even more stunning than the small minivet is the **Long-billed Minivet** (20 cm) (*Pericrocotus euthologus*), dramatic in scarlet and black, and a winter visitor to the capital. I have seen them on the northern Ridge and at Sultanpur and Buddha Jayanti Park. The females are equally dramatic in yellow and black. The birds utter a pleasant double-noted whistle 'Tui-tui!' as they flit among treetops.

Whitebrowed Fantail, *Rhipidura aureola*, Chak Dil

This delightful little (18 cm) sooty-brown bird with its broad white forehead and 'eyebrow' (supercillium) usually draws attention to itself as it prances and pirouettes among the branches, bowing and curtseying—all the while fanning open its white-tipped brown tail. Normally found in pairs, whitebrowed fantails are unfortunately rarely encountered residents of Delhi. I have seen them very sporadically on the northern Ridge, and at

Sultanpur. Look hard in similar habitats elsewhere and you may well be rewarded!

The birds have a harsh 'Chuk-chuk!' call, but a sweet 'Chee-chee-cheeweechee-wi!' song, sung as the bird dances around. They nest chiefly between March and June.

Black Drongo, *Dicrurus macrocercus*, Kotwal

A slim, alert, fork-tailed glossy black bird (18 cm) that is a familiar sight perched on transmission wires overlooking fields and open areas in and around the capital. Drongos can also be met with in most of Delhi's large parks and open spaces, such as Buddha Jayanti Park, Lodi Gardens, Deer Park, both the Ridges, the gardens around Humayun's Tomb, the golf course, and the zoo.

Members of the crow tribe, drongos are pugnacious in the defence of their young and opportunistic enough to flutter low over a field that is being ploughed up, or stubble that is being burned, to snap up fleeing (and semi-tandoori) insects and small animals. They are remarkable mimics of other birds and can bamboozle you easily, though they do have a seemingly typical fluty call of their own. Drongos nest between April and August, laying three to five whitish eggs (with reddish spots) in a flimsy-floored, cup-shaped nest made out of twiglets and fibres bound up by cobweb.

Several years ago, in winter, I was introduced to a solitary **Spangled Drongo** (*Dicrurus hottentottus*) in Deer Park at Hauz Khas. With its glossy gelled plumage and pop star hair-like crest, it sat somewhat stolidly in a eucalyptus grove, fending off the unwelcoming house crows. I haven't been back to check it out since.

Asian Paradise Flycatcher, *Terpsiphone paradisi*, Doodhraj

Spotting a paradise flycatcher in Delhi will easily be the highlight of any birding trip and your best chances to do so are between March and April, and then again between August and October when the birds pass through Delhi en route to and from mysterious nesting destinations.

The male (20 cm) is satin white, except for a blue-black, raffishly crested head, a somewhat large bristle-moustached bill and neon-ringed eyes, and soft dove grey to black 'racing lines' where his wings fold into his body. His *pièce de resistance*, of course, is his glittering, 35-cm-long, white twin tail ribbons that he flutters behind him like an expert rhythmic dancer. The female, and males under three years old, have the same blue-black heads, but are autumn chestnut on top and off-white below. Females don't have the tail ribbons either.

I have seen these glorious birds on the northern Ridge and at Buddha Jayanti Park and Sultanpur in spring—which is the best time to see them. I have also regularly been meeting them on my walks on the northern Ridge between July and October. But by this time, the males have dropped their glamorous ribbons and don't look nearly as exciting. The birds often give away their presence by their harsh, abrupt-sounding 'Chwit! Chwit!' calls, though they do have a pleasing seductive song while courting. Paradise flycatchers eat insects, naturally specializing in flies, midges, and the like.

Blue-capped Rock Thrush, *Monticola cinclorhynchus*, No local name

Sometimes you can get lucky—even on a birding trip to an ordinary park in Delhi. Peering into the lightly wooded forest of the New Delhi Ridge from the edge of Buddha Jayanti Park in April 2000, for an elusive paradise flycatcher, I spotted something even more exciting.

A small (18 cm) blue and chestnut bird with a flashy white wing patch was sitting quietly on a semi-sunlit branch. Its head and chin and throat were a brilliant cobalt blue, the rest of its upper parts dark inky blue black. Its rump and breast shone bright chestnut and the white splotch on its wing was quite prominent. I was lucky to be able to snap off a couple of pictures before it flew off further into the interior of the Ridge. It was, I found out later, the blue-capped rock thrush, a bird that has been listed as 'accidental' for Delhi. A summer dweller of the Himalayas, blue-capped rock thrushes spend the winter on the subcontinent and are especially fond of the Western Ghats. The bird I saw was probably en route to its home in the mountains. It was (luckily for me) a male: females are plain brown on top with a scaly patterned brown and white underside and barred rump.

Birds are creatures of habit and I would not be very surprised if the blue-capped rock thrush were to be regularly spending some time on the New Delhi Ridge every year—on its way to or from the mountains.

Red-throated Flycatcher, *Ficedula parva*, Turra

This small (11.5–12.5 cm), jumpy brown flycatcher is fairly common in Delhi's parks and gardens especially during October and November and then again during March and April, when it passes through to and from its home in the Himalayas to its winter retreats further south on the subcontinent. Some birds spend the winter in Delhi, flitting and tossing themselves after tiny insects in woodland areas, parks, and gardens all over the capital.

Round-headed and round-eyed (a white ring around the dark eye accentuates the wide-eyed look), these droopy-winged birds give their presence away by their harsh 'Trrr!' calls, that sound somewhat like a defective electric bell. A tomato-orange splotch on the breast distinguishes males from females. The birds constantly jerk up their tails as if to show off their neat white rumps.

Grey-headed Canary Flycatcher, *Culicicapa ceylonensis*, Zird-phutki

This diminutive (12.5 cm) ash-grey and sulphur-yellow flycatcher is a regular winter visitor to areas like the northern Ridge, where

I have been seeing it consistently over several years. Restless and impatient, it seems to prefer areas near water, and perches on overhanging branches, rocketing out of cover to zap up an insect or charge a startled bulbul. Grey-headed canary flycatchers appear to be happy in Delhi too, as they can be heard singing their cheery five-noted 'Chik-whichee-wichee!' songs quite frequently, which is a sound familiar to those who go to the Himalayan hill stations in summer. I have also seem them at Sultanpur. Other likely places would be Lodi Gardens, the gardens around Humayun's Tomb, Deer Park, and Buddha Jayanti Park. The birds breed in the Himalayas.

Bluethroat, *Luscinia svecica,* No local name

A small (15 cm), sprightly earthy-brown 'runabout' bird, with a pale buff eyebrow and a sky-blue breast marked with a red bull's-eye spot and graced with a double gorget of chestnut and black. The female lacks the chestnut and blue, but the black-tipped auburn tail, conspicuous in flight, is common to both sexes.

The bluethroat is a rather unobtrusive silent winter visitor to Delhi, arriving by around mid-September and staying on till perhaps early May. It is fond of damp and dank areas, and runs around the edges of pools, ponds, swampy tracts and such places, picking up insects.

I have seen bluethroats along the banks of the Jamuna, on the northern Ridge, at Sultanpur, and at Lodi Gardens. Veteran Delhi birders maintain that the number of bluethroats has drastically declined in recent years.

Oriental Magpie Robin (Magpie Robin), *Copsychus saularis,* Dhayal

Easily Delhi's best songbird, the magpie robin, smart and diffident in its glossy tuxedo, begins its beautiful recitals around February–March, singing with great verve and sweetness from the tops of exposed perches in parks and gardens all over the capital. The male, who sings, has a glossy black head, breast, back and wings, counter-matched by a shell-white underside and white flash across the wing. The tail, often cocked jauntily over the back, is black and white. The female is sooty brownish black on top and ashy grey on the chin and, like the male, white elsewhere.

Interestingly, the magpie robin had been first listed as a winter visitor to the capital. In the last fifty years, however, it has obviously made Delhi its home and nearly every park and garden will have at least one pair in attendance. Magpie robins are pugnaciously territorial birds, and in March and April you may come across males chasing each other furiously through the canopy, pausing only to defiantly belt out a few bars of song as they try to wrest control of an area from one another. (A sight which always reminds me of highly strung classical music maestros in tuxes and tails chasing each other around an auditorium.) Once territorial matters are sorted out, pairs nest (the season lasts from April to July) in the holes of walls or tree trunks, stuffing these with grass rootlets and soft rubbish. Between three and five light blue-green eggs, splotched in rust, are laid, and while the female incubates, both parents feed the hatchlings. Often two broods may be raised.

Magpie robins eat insects, and during the early morning or late evening may be heard to utter a long plaintive 'Swee-ee!' whistle, or harsh 'Chrr-rr!' usually from the depths of foliage. This is just a 'keeping-in-touch' call. They are also good mimics of other birds.

Indian Robin, *Saxicoloides fulicata*, Kalchuri

A close country cousin of the magpie robin, the Indian robin (18 cm) is perhaps as engaging in its own territory—preferring the drier, boulder-strewn areas of parks and the Ridge. The male is a matt dried-leaf brown on top, with a glossy black breast and tail and white wing patch. The tail is cheerfully cocked most times to show off an attractive autumn-leaf rump. The female is duller, ashy brown, and does not have the wing patch.

Indian robins sing in January and February with the male uttering a brief series of high-pitched rapid warbles. The birds nest between April and June, putting together a pad of grass, rootlets, and rubbish, lined with feathers under a stone or in a low-placed hole in a tree trunk or earth bank. Two to three green-tinged creamy-white eggs, splotched with rust are laid and both parents bring up the chicks. The birds eat spiders and insects.

The rocky areas around Badkhal Lake abound in Indian robins and you are almost sure to see them in the boulder-strewn sections of Buddha Jayanti Park and both Ridges.

Black Redstart, *Phoenicurus ochruros*, Thirthiri

A neat, grave winter visitor to gardens and parks throughout the capital, the black redstart (15 cm) will enchant you with its immaculate manners. Constantly bowing and shivering its autumn-tinted tail, this upright-standing black and chestnut bird (with a cigarette-ash crown) arrives punctually in favourite gardens and parks around the middle of September and will stay on till the end of April. The female is brown and duller, but has the same rust-orange tail.

On arrival redstarts immediately stake out territories and will defend these determinedly from one another. Sitting quietly on a low branch or boulder, they plane down to pick up insects from the ground, their rusty tails flaring alluringly. They also like the rock-strewn open areas of the Ridge, for example, as well as thorny acacia copses. They may also use certain areas (like my garden for instance) as stopover or transit points en route to their final wintering destinations. There have been fears in recent years that the numbers of redstarts visiting Delhi has fallen. The birds are generally silent.

Common Stonechat (Collared Bushchat), *Saxicola torquata*, No local name

A small (17 cm) attractive bird, peppercorn black above and

rufous ochre below, with a broad white collar and streak on the wings. The female is brown streaked above and ochre rufous below.

Stonechats are winter visitors to the open country, agricultural fields, and beds in and around Delhi, where they perch on tall grass stems, boulders, or bushes, looking out for insects on the ground, flicking their tails up and down. I have seen them at Sultanpur and in marigold and mustard fields nearby, and on the cattail fringed banks of the Jamuna at Okhla. The birds arrive from the Himalayas in September and leave by the end of April. The numbers may increase in around March, as those that winter further south begin their northward journey and pass through Delhi.

Pied Bushchat, *Saxicola caprata*, Kala Pidha

A small (13.5 cm), dumpy, round-headed black and white bird that often gets mistaken for the magpie robin. The pied bushchat is almost entirely black with white on the rump, wings and abdomen, and, unlike the magpie robin, is found in open country, scrub areas, fields and wastelands, clinging to tall grass stems and reeds, looking out for insects. The female, usually seen together with the male (but a sensible distance away), is pale brown, tinged with rufous on the rump.

Pied bushchats are residents; the males start singing (with moderate talent) their 'Chip-chipee-chiwee-chrr!' song in January and nesting commences in the following month, the season lasting till perhaps August. A cup-shaped nest is constructed on the ground under a tuft of grass, or in a hole in a wall or the earth bank. Between three and five pale bluish-white eggs, speckled with reddish brown, are incubated solely by the female, though both parents bring up the brood. It is thought that the parents take their fledglings (peppery-brown and spotted and speckled and very woebegone little things) into reed beds and clumps of grass, to teach them how to forage, and I have on several occasions come across these unhappy little toddlers squatting miserably in the thickets, alone in the world.

Indian Chat (Brown Rock Chat), *Cercomela fusca*, No local name

A small (17 cm), rather plain brown bird with darker wings and a blackish tail that is often slowly and thoughtfully raised and lowered. The Indian chat loves old forts and abandoned monuments—of which there are plenty in Delhi. I have seen the birds at Tughlaqabad, the walls around Isa Khan's tomb, the ruins of Feroze Shah Kotla, and the Jamali Kamali—to name a few such places. The birds are residents of such areas and nest between April and August, fashioning a small cup of straw, rootlets, and stalks in a hole in a wall or suitable niche or cleft. Three to four pale blue eggs, with rusty specks and spots are laid. The male sings a sweet thrush-like song, but the birds are otherwise silent.

Brahminy Starling (Black-headed Myna), *Sturnus pagodarum*, Brahmini Myna

This cocky, self-assured grey and fawn myna (21 cm) has a glossy jet-black head with a recumbent, somewhat undisciplined crest which is hilariously raised when the bird gets excited or has thrown back its head to sing and woo its beloved. The tail is brown with a frilly white edge, the madly staring eyes are whitish, the bill tipped bluish.

Brahminy starlings can be found sauntering around in pairs in gardens and in small groups in open scrub country and wasteland furnished with bushes and trees. Buddha Jayanti Park, Delhi zoo, Asola sanctuary are such favourite haunts, but you can with equal certainty meet them in your garden or neighbourhood park.

The male plunges whole-heartedly into the courting ritual in around March, flinging back his head, rolling his eyes dementedly, raising his pop-singer's crest, and singing his song (far superior to that of the average pop singer) with the passion one usually encounters only in Hindi films. The birds also converse in a pleasant chattering, chortling dialect. For nesting, they stuff grass and rubbish into a hole in a tree trunk or wall, where three to four unsullied pale blue eggs are deposited. Both parents raise the brood.

Rosy Starling (Rosy Pastor), *Sturnus roseus*, Tilyer

Pale pink and glossy black, the rosy starling is a passage migrant through Delhi and is best spotted between July and August and then again between March and April, en route to and from its wintering grounds. The wings and shaggy-looking head are black, the rest of the body is rose pink. Some birds possibly do remain in the Delhi area through the winter, and may be seen in open cultivated areas, garbage dumps and fields, usually in the company of bank mynas and starlings. They feed on insects, grain berries, and nectar. I have seen large flocks roosting at Sultanpur, and one evening in April, I watched squadron after squadron, fly northwards over my house, probably on their way back home in Central Asia.

Common Starling, *Sturnus vulgaris*, Tila Maina

This slim, hyper, glossy black myna (23 cm), iridescent in metallic green, purple, and blue tints, has its hackled plumage stippled with ivory, making it look from a distance as though it has been studded with cumin seed. A winter visitor, it arrives by around mid-October and leaves by the end of March. It keeps company with such local relatives as pied starlings and Indian mynas, and can be seen sauntering confidently around garbage dumps, stubble fields, and swampy areas in and around Delhi, as well as tree-shaded parks and gardens.

Common starlings, with their restless type-A personalities, are entertaining and hilarious to watch, though some of their dietary preferences (excreta!) can be rather off-putting! They do appreciate fruit too, which is some saving grace. By around the end of March, noisy rabbles collect in trees and on transmission lines prior to migration, like conventions of highly strung investment bankers watching a stock-market crash.

Asian Pied Starling (Pied Myna), *Sturnus contra*, Ablak Maina

A somewhat rotund-looking black and white myna (23cm) with an orange-juice-coloured naked patch of skin in front of the eyes and similarly coloured eyelid and bill. The breast and chin and upper plumage are black, the rest of the underparts (and an ear patch) are white. The legs are yellow.

Pied starlings like well-watered lawns (like other mynas they love bathing) and any area where there is water nearby—the fields and country beside the Jamnua, Sultanpur jheel, and such like—is sure to have them around. I have seen them come to roost at Sultanpur in huge screeching flocks. They are chiefly insectivorous. They have a pleasant musical call and nesting commences in May and continues till August. An untidy dome-like structure is constructed out of sticks and twigs and stuffed with rags and straw, some way up a tree—usually close to water—or even in a set of traffic lights as I once saw near Rajghat. Three or four families may nest in close proximity. Three to four glossy blue eggs are laid and both parents share all domestic chores.

Indian Myna, *Acridotheres tristis*, Desi Myna

This familiar coffee-brown myna (25 cm) with its glossy black head, taxicab-yellow eye-patch (and beak and legs), and white wing patch, really needs no introduction. Garrulous, self-assured, and swaggering everywhere with city-bred savoir-faire, this myna likes nothing better than a good dust-up (with an appreciatively shrieking audience) WWF-style, or long blissful afternoons of flirting and chortling tenderly with its beloved, nodding its head and frowzling up its crest with comic ceremony as it attempts to flatter and seduce.

Nesting begins in April and holes in tree trunks and walls are stuffed with rags and straw and such suitable furnishings. Four to five glossy blue eggs are laid. Mynas are omnivorous and great shoplifters, sauntering around market places and eating houses, filching fare with great nonchalance and élan—but with one eye always open to danger. Grasshoppers and the like too are dispatched with ease. A cat or an owl, if spotted, will be mercilessly heckled and harassed, and a backup mob of other species—babblers, crows, and sparrows —will quickly join in to assist them drive the beast away.

Bank Myna, *Acridotheres ginginianus*, Ganga Myna

Slightly smaller than the Indian myna at 23 cm, the bank myna is clad in thundercloud grey with bright brick-red skin around the eyes and matching orange beak. Its wing patch is off-white.

Bank mynas are quite common in Delhi and like Indian mynas are great market-place aficionados. They also seem to love sauntering about *dhaba*s and open-air *chai-dukan*s, picking up whatever comes their way. Flocks also congregate in fields, they ride bareback on rhinos and nilgais at the zoo, they examine luggage at railway stations, and, at the end of the working day, repair to roosts—in acacia groves for instance—where they settle down for the night after their usual raucous parliamentary debating sessions.

Bank mynas start nesting in March and may continue to do so till August. Weep holes in walls—and even those in the sides of Delhi's flyovers—make favourite sites for apartment colonies, after being suitably furnished with grass, rubbish, and feathers. Earthen banks are also regarded as hot colony sites. Between three and five glossy pale-blue eggs are laid, and the fledglings, usually seen by July and August, look like unwashed, uncivilized versions of their parents. They appear to behave that way too—on several occasions, I have seen gangs of young hoodlum mynas aggressively pursue an adult in the hope of a handout.

Martins

Martins, closely related to swallows, are swift, fastback-winged birds, always scissoring across the skies beaks agape as they trawl tiny insects. They are often frustrating to try and identify and you need patience and deft tracking abilities to follow them in flight through your binoculars.

The **Plain Martin**, ealier the Sand Martin (*Riparia paludicola*, Ababil), is a small (10.5 cm) dusky brown bird, pale and smoky on its breast and whitish underneath, with a slightly forked tail. Flocks of these birds can often be seen flying over the river, and sometimes great numbers line up on transmission wires nearby. They are residents and nest between October and March, excavating tunnels in the sandbanks along the river.

The **Dusky Crag Martin** (*Hirundo concolor*, Chatan ababil), just a little bigger at 13cm, is dark sooty brown with white spots on all of its tail feathers save the central and outermost pair. Small numbers of these birds may be seen around the ruins of old tombs and monuments, such as Humayun's Tomb and Purana Quila. They nest between June and October, fashioning a deep oval saucer out of mud, and stick it to a suitable wall, or under an arch or parapet. Two or three white eggs with reddish specks are laid, and both parents look after the brood.

Swallows

Similar to martins and swifts, swallows too can be frustratingly difficult to tell apart in the field, as they flit and flicker swiftly across the sky, twisting and turning at all the wrong moments and at incredible speeds.

The **Barn Swallow** (*Hirundo rustica*, Ababil), till lately known as the Common Swallow, is a winter visitor to the Delhi area, arriving by November and collecting in great numbers on transmission wires in April, prior to its departure. An attractive steel-blue bird (18 cm) with a shiny chestnut forehead and chin, steel-blue gorget, and creamish under parts, this swallow has a deeply forked tail, necklaced with white. It is common over, and around, the Jamuna, and may even patrol the moats at Delhi zoo, which enables you to get a good close look at it. Sultanpur and Badkhal are two other favourite areas, and, at the latter, I have seen masses of them lining up on the transmission lines at the end of March.

The smaller (13.5 cm), natty **Wire-tailed Swallow** (*Hirundo smithii*) is glossy blue above and white below, with a chestnut cap and two antennae-like wires sticking out of its tail. You can often meet them at the river, and around other water-bodies. They nest between February and August, constructing saucer-like homes out of mud, in the cornices of structures. Three to five small white eggs marked with reddish specks are laid, and often two broods are raised. The area around Sultanpur jheel appears to receive an influx of wire-tailed swallows in June and July, and the birds do nest in the area: I have seen their nests inside abandoned tube-well shelters and under culverts there.

The **Streak-throated Swallow** (*Hirundo fluvicola*, Nahar ababil), better known perhaps as the Cliff Swallow, at 11.5 cm, is another tiny member of the fraternity. Glossy steel blue above, it has a pale-brown rump and dull rusty forehead. The underparts are off-white, profusely stippled with brown on the throat.

A resident swallow, the bird nests in colonies, dabbing together untidy clay-pot beehive-like structures under bridges and culverts or gateway arches. Three to four white eggs are usually laid.

The **Red-rumped Swallow** (*Hirundo daurica*, Masjid Ababeel) is the largest (20–3 cm) and slowest-flying of swallows, preferring to hawk insects around monuments and ruins, usually well away from water (though I have seen them at Sultanpur jheel). Glossy steel blue above, off-white below, its chestnut half-collar on the back of its neck and chestnut rump (especially visible when it banks) are pointers to its identity. Its tail is deeply forked. The nesting season is from March to September and a small circular chamber with a flask-shaped entrance is fashioned out of mud pellets and stuck to the underside of bridges, culverts, and ruins

Bulbuls

Ever-cheerful residents of Delhi, bulbuls can be found in gardens, parks, woodland areas, rocky scrub country (like that of the Ridge and the Jawaharlal Nehru University campus), avenue trees, and just about any green or not-so-green patch in the capital.

The commonest by far is the sooty brown **Red-vented Bulbul** (*Pycnonotus cafer*, Bulbul), a pugnacious, no-nonsense bulbul (20 cm), with a black head and truncated crest, coffee-brown body scalloped (especially about the neck) with off-white, an off-white rump and vivid scarlet vent. Like so many of the capital's citizens, the red-vented bulbul is always ready for a punch-up or dust-up, and during the breeding season (March to August) is known to gang together and attempt to lynch owls.

The more ceremonial and 'posh'-looking **Red-whiskered Bulbul** (*Pycnonotus jocosus*, Sipahi Bulbul) is about the same size, but looks slimmer and holds itself more erect and guardsman-like. With its jaunty black crest, bright scarlet cheek patches (that from close up look like tufts of scarlet hair sticking out of the ears!) and black 'chinstrap' markings, this beige-brown (on top) and white (below) bulbul looks as though it would be more at home doing ceremonial goose-steps in the forecourt of Rashtrapati Bhavan, than being bullied in the streets by those vulgar red-vented Bulbuls. It too has an orange-crimson vent. Red-whiskered bulbuls are also perhaps the most cheerful and musical of Delhi's bulbuls.

The silver-brown (on top) and white (below) **White-eared Bulbul** (White-cheeked Bulbul, *Pycnonotus leucotis*) has a black head, with a 'crest' that looks as though it has just been chopped off, and a gleaming white cheek patch. Its vent is a cheerful sunny yellow. Not as common as the other two, you can still almost always see them in places like Buddha Jayanti Park and the campus of Jawaharlal Nehru University.

Bulbuls eat insects, fruit, flower petals (and so can be destructive in gardens), vegetables and grain, and will often gather at feeding spots in parks and gardens all over the capital. Nesting commences in March, lasting perhaps till August, the prime months being from June onwards. The female builds a neat shallow cup out of fine twigs and grasses and lines it with rootlets and hair, placing it usually in a bush, shrub, hedge, or tree. Three or four pinkish eggs are laid and both parents look after the brood. It is well worth keeping a watch on your garden hedge during the bulbuls' breeding season: a pair may take up residence in its dim greeny depths and raise their family there.

Prinias

Prinias, better known perhaps as wren-warblers, are small (12–13 cm), slim, long-tailed birds accomplished at skulking unobtrusively at the bottom of hedges or flitting exasperatedly from branch to branch, before you can see them properly. There are several species to be found in Delhi, of which we shall met three here—two of the commonest and the third, amongst the most beautiful.

The **Plain Prinia**, till lately the Indian Wren-Warbler (*Prinia inorata*, Phutki), is a tiny, long-tailed, dun-coloured warbler, quite nondescript really, with off-white underparts, eyebrows and eye-ring, and a black bill. In winter, the bird appears more rufous and the bill turns brown. Fond of open agricultural country where the crops stand tall, as well as grassy thickets, bushes, and spiky acacias, plain prinias can be commonly found in the areas along the river south of Okhla, parks such as Buddha Jayanti Park, both sections of the Ridge, and indeed any garden with bushes and hedges. The birds start singing in around February, with the males clinging excitedly to tall grass stems and belting out their incessant 'Tleek-tleek-tleek!' song at considerable volume (in relation to their size). Nesting takes place between March and September.

A pouch of fine woven grass is intertwined with a tussock of grass and three to five glossy greenish-blue eggs, marked with reddish brown, are laid.

The **Ashy Prinia** (Ashy Wren-Warbler, *Prinia socialis*, Kali Phutki) is a familiar resident of probably every garden hedge in Delhi. Sleek looking, loosely put together, and long tailed, it is steely grey above, fulvous white below, with staring orange eyes and a tail that is constantly being flicked up and down. During winter its plumage turns deep reddish brown.

The ashy prinia loves skulking within—and at the bottom of—garden hedges, picking up insects, letting off its warning 'Tee-tee-tee!' call if disturbed. While nesting lasts from March to September, the monsoons are the favoured period. The ashy prinia may emulate the famous tailor bird and make its home in a funnel of stitched-together leaves or, alternatively, weave a delicate oblong purse of fibres, using cobweb to hold it together, and attach it to the supporting leaves of a bush. Three or four glossy brick red eggs are laid and both parents look after the brood.

The **Yellow-bellied Prinia** (*Prinia flaviventris*) is smartly turned out with an ashy-grey crown, thin white eye-stripe, buttercup-yellow belly and olive-green wash to its upper parts. (The first time I saw it—at Isa Khan's Tomb—I thought it was an exceptionally glamorous ashy prinia). The areas along the river, where grass tufts and cattail spring high, are likely places to see this warbler. Yellow-bellied prinias nest between April and October, constructing oval-shaped nests out of fine grass, fixed to grass stems or in bushes. They have a pleasant, trilling five-noted song.

Prinias snap up small insects, caterpillars and the like, and also help themselves to nectar.

Oriental White-eye, *Zosterops palpebrosus*, Baboona

A tiny (10 cm), solemn, goggle-eyed greenish-yellow and off-white bird that jingles softly in the canopy with others of its kind, the white-eye is a common arboreal resident and can be quite difficult to spot.

Somewhat dumpy and square-tailed, the white-eye's dark eye is ringed with white feathers; its breast and throat are bright yellow, the belly dirty white. Fond of berries and nectar, white-eyes jingle through the canopy in search of these, in parks and gardens throughout Delhi. Flowering trees and bushes—like the bottlebrush or honeysuckle or poinsettia—are magnets to these birds. Any neem tree with berries is sure to have a small flock in attendance.

The birds nest between March and August, with both partners skilfully constructing a tiny, lovely cup of fibres, bound together with cobweb. Two or three pale blue eggs are laid. It is often easier to detect the presence of white-eyes by listening for their soft, musical, jingling calls.

Blyth's Reed Warbler, *Acrocephalus dumetorum,* Podana/Tiktiki

One of the vast, exasperating buff and dun warbler clan, the Blyth's reed warbler (14 cm) is a passage migrant through Delhi, seen in August–September and then again between March and June.

Olive brown above, with a white throat and buff underparts, this warbler has a longish bill and a barely discernible white eyebrow or supercillium. It tends to spread out its tail in flight. It is fairly common in parks, gardens, groves of acacia trees, as well as bushes in swampy and riverine areas. It hops and flits amongst the bushes, occasionally uttering a single 'Chuk!'. The birds are known to sing while on their spring migration.

Clamorous Reed Warbler (Indian Great Reed Warbler), *Acrocephalus stentoreus,* No local name

A well-named giant (19–20 cm) amongst warblers, the clamorous reed warbler is easier detected by the noise it makes—a loud 'Karra-karra-kareet-kareet!'—as it skulks amidst the reed beds of the Jamuna and countryside around Okhla, than by sighting. Look hard at from where the racket comes and you may see a furtive pale olive-brown warbler (whopping large by warbler standards, but otherwise warbler-like in shape and colour), with a large stout bill and (usually) open orange mouth, flitting dexterously amidst the reed stems like a hawker in a bus.

These birds are thought to be residents as well as passage migrants, thronging the reedy areas of the river in April and May. Resident birds nest between May and August. They build a deep and large cup out of dry reed leaves, and attach it between reed stems. Three to six pale green or greyish-white eggs are laid and both parents look after the brood.

Lesser Whitethroat, *Sylvia curruca*, No local name

Lesser whitethroats appear not to approve of birders. They flit about in the canopy making disapproving 'Tch-tch!' noises as you peer up through the foliage trying to spot them. If you are persistent you may spot a small (12 cm) earthy-brown (above) and creamy-white (below) bird, with a grey cap and white cheeks hopping and scuttling amongst the branches, and leaning over alarmingly from time to time to pick an almost-out-of-reach berry or caterpillar. (Alas, it never does actually lose its balance and topple over!) Lesser whitethroats are migratory and arrive in Delhi's gardens and parks and woodlands by around the end of September, staying on till the end of April. They eat insects, caterpillars, berries, and are also fond of nectar.

Tailor Bird, *Orthotomus sutorius*, Darzee, Phutki

A cheerful tiny-tot (12.5 cm) warbler, olive green on top and off-white below, with a rusty cap and a long pin-pointed tail which is flicked jauntily over its back, the tailor bird flits amidst bushes and creepers and hedges, shouting 'Towit-towit-towit!' or 'Pichick-pichick-pichick!' at the top of its voice. Common throughout Delhi, in gardens and parks and woodland areas, the tailor bird is renowned for the way in which it stitches leaves together to form a funnel in which it places its nest, made out of fibres, feathers, cotton waste, and the like. Three or four pale pink or greenish eggs, boldly blotched with brown or black are laid. The nesting season lasts from June to August and the nests are usually found low down in broad-leaved bushes or shrubs.

Tailor birds eat tiny insects and grubs, and are also fond of nectar. They are cheerful even if loud companions in every garden.

Greenish Warbler (Greenish Leaf Warbler), *Phylloscopus trochiloides*, No local name

A small (10.12 cm) dun olive-green warbler, with a pale yellow eyebrow (or supercillium) and dark streak through the eyes. The rump may be a brighter yellow-green and a whitish wing bar may be present. The tail is brown.

A passage migrant, the bird is found commonly in gardens and parks, emitting a 'Tysip!' call note every now and then. The birds pass through from August to October, but have apparently not been noticed on their way back in spring.

Chiffchaff, *Phylloscopus collybita*, No local name

A tiny (10.5 cm) dusky, sooty warbler, with a moss green tinge to its plumage, dirty white below and with a restless wing-flicking habit. The legs are black, the eyebrow pale, and there are no wing bars.

A fairly common and quiet winter visitor, the chiffchaff lurks in the canopy or in bushes and reed beds picking off tiny insects. Once, possibly in March sometime, on the northern Ridge, I was puzzling over the identity over these birds, when I heard one of them utter softly 'Chiffchaff, Chiffchaff!' and the mystery was solved! The birds arrive in September and stay on till April.

Babblers

In their dun, earthy hues, babblers may not be the most colourful of Delhi's birds, but what they lack in glamour they more than make up for in personality and charm. They are, by and large, untidily plumaged brown birds, with glowering eyes and loose dangly tails, and appear to be quite hopeless at flying. They flutter and glide desperately, always looking as though they're about the crash land. Babblers like each others' company and will scout around the bottom of garden hedges, turning up the dirt and detritus for spiders, cockroaches, and other such delicious insect fare. However, they also relish nectar, berries, and fruit. They build (usually) neat cup-shaped nests out of grass and rootlets and leaves, installed in bushes, hedges or trees. Despite their ferocious expression and harsh language, babblers are really quite foolish and soft in the head. Their nests are regularly parasitized by hawk-cuckoos and pied-cuckoos.

The **Common Babbler** (*Turdoides caudatus*, Dumri Penga), at 23 cm, is not as common as its name perhaps suggests. It is a heavily streaked earth-brown babbler with a slim long tail (trailing behind it on the ground as it hops about), pale tawny undersides and creamish-white chin and breast. It scuttles mouse-like, or bounces along in kangaroo fashion, along the bottom of garden hedges, in small groups. Other habitats preferred include open cultivated country, parks and the rocky wastelands around Delhi, and, of course, both sections of the Ridge. The common babbler also likes perching prominently on the tops of bare bushes or perches, from where it lets forth its alarm-clock-like musical trill. It nests between March and September, and three or four glossy turquoise eggs are laid. Both parents bring up the brood.

The **Striated Babbler** (*Turdoides earlei*, Bara Penga) at 25.5 cm, is a babbler that favours the swampy, riverine tracts along the Jamuna and Okhla. A little larger and bigger-built than the common babbler, it too is earthy brown and heavily streaked along the head and back, as well as the throat and breast. Its call is a musical 'Tew-tew-tew!' uttered feelingly from the tops of tall grasses or reeds. It nests twice without the year, raising two broods, the first between March and May and the next between August and October. The nest is a large cup-shaped structure of leaves, grasses, and rootlets, affixed to stems or placed in a bush. Three or four pale blue eggs are laid.

The **Large Grey Babbler** (*Turdoides malcolmi*, Sat Bhai) at 28 cm, is a sandy-grey babbler with an ashy rump and ash 'caste' marks on the forehead, and white outer feathers to its tail. It canters about in flocks of a dozen or more, head held high, in parks, areas of open cultivation, and along Delhi's many tree-lined avenues, glaring disapprovingly out of its cold, pale yellow eyes. Perceived threats (or anything else causing excitement or disagreement) are greeted by the whole flock insanely uttering 'Kay-kay-kay-kay!' in loud machine-like tones, and frowzling up their plumage, drooping their wings and loosely jerking their tails, as if to frighten off the threat by this display of collective and ferocious insanity.

Large grey babblers appear to have become less common in Delhi in the past twenty-odd years. You can, however, usually meet them in the groves of Delhi zoo, Lodi Gardens, Deer Park, Buddha Jayanti Park, both sections of the Ridge, the woodlands of Sultanpur, and any well-shaded local park.

The **Jungle Babbler** (*Turdoides striatus*, Sat Bhai) at 25 cm, is by far Delhi's commonest babbler. 'Sisterhoods' of six or seven of these glowering khaki-brown birds, with their enchanting frowns and gimlet-yellow eyes and bills, are found in every garden, busy turning up the dirt at the bottom of hedges, muttering darkly amongst themselves and quarrelling raucously every now and then. They seem to be insatiably curious and will turn over anything—bottle caps, pieces of paper, empty cigarette packets—to see what treasures or scandals lurk beneath.

Jungle babblers seem to adore grooming each other and being groomed. They will huddle ecstatically together and ruffle through each other's plumage, eyes closed in bliss, as all the right spots are duly massaged. They breed between March and September, and, apparently, several birds take part in nest building, feeding and care of the young. Inevitably, these hard-eyed softies fall victim to the wiles of pied cuckoos and bring up their giant youngsters—something that causes them quite a lot of stress, when the giant youngster outgrows them hugely but still behaves like a baby.

The **Yellow-eyed Babbler** (*Chrysomma sinense*, Gulal-chasm) at 18 cm, looks more like an outsized warbler than a babbler. Somewhat round headed, this attractive chestnut-brown (above) and cream-white (below) babbler has virulent orange rings around its eyes, which are far more prominent than the yellow iris that has given it its name. Found in small groups in thorny scrub country, or amidst thickets of tall grass (the country south of Okhla is a good area), the birds, in the breeding season, clamber up to exposed perches and trill their 'Trit...trit...rit...rit!' song from these. Usually they cling and clamber about within the thickets and grass clumps, and are therefore difficult to see clearly. The birds nest between June and September. A deep, tidily made cup of grass, plastered with cobweb is placed in bushes and four or five yellowish-white eggs, freckled with purplish brown, are laid.

Larks

Larks are dun, sandy little sparrow-like birds that favour open country, and can be met in the vast thorny scrub country of the Ridge, the areas around the river, and in cultivated fields and wastelands around Delhi. They are known for their flight display and song and some performances can be truly breathtaking. They breed in spring and summer (usually between February and June) and are, by and large, ground nesters. A cut-like depression in the ground, lined with hair or grass often suffices as a home. Between two and four eggs may be laid, depending on the species, usually brown-splotched yellowish white or greyish white in colour. Larks are seed and insect eaters.

The **Indian Bushlark** (*Mirafra erythoptera*, Junghi Aggia), till recently known as the Red-winged Bushlark, is a small (14 cm) brown lark, well streaked above, pale brown below; and with bold spots on its breast. A rusty patch on its wings is visible both when the bird is perching as well as in flight. This lark prefers dry, thorny scrubby areas as well as cultivated fields. Singing and displaying begins in February. The male flutters straight up from his bush-top perch, 10 m or so, emitting a squeaking 'Si-si-si-si!', which quickly gives way to an equally high-pitched 'Wisee-wisee!' that slows down in tempo as the bird runs out of steam and planes back to its perch on stiffly outspread wings.

The **Greater Short-toed Lark** (*Callendrella brachydactyla*, Pulluck) at 14 cm, is a winter visitor, always seen in large busybody groups in open country, cultivated fields, and wastelands. They are sandy brown above, stippled with black and off-white below, and with a white eyebrow (supercillium).

The **Indian Short-toed Lark** (*Calandra raytal*, Retal), till recently known as the Sand Lark, is the smallest (12 cm) and most washed-out looking of larks. It is pale ashy-sandy above, faintly streaked with black, dull white below, and has fine stippling on the breast. The bird looks as though someone had given it the once over with a pencil eraser!

This lark is a fairly common resident and stays around the areas near the river, merging uncannily into the ground when it keeps still. The male displays by flying up some 30 m or so, fluttering and gliding by turn, singing all the while, before rocketing back to earth and levelling out just over ground zero.

The **Crested Lark** (*Galerida cristata*, Ghendul) at 18 cm, is the largest of the larks found in the area, and is unmistakable because of its erect, pointed crest. It is sandy brown on top, streaked with black, and has a pale eye stripe. The tail is dark brown and whitish on its outermost feathers.

The crested lark is found in typical lark country—from the riverine tracts to dry boulder country. It displays in the usual lark manner—flying high and singing, then dropping to earth and disappearing amongst the clods.

Perhaps the award for the 'best performer' amongst the larks should go to the inconspicuous-looking **Oriental Skylark** (*Alauda galgula*, Bharat), till recently known as the Little Skylark. Quite like a female sparrow, the Oriental Skylark (16 cm) has a short truncated crest and is heavily streaked in dark and light brown above, pale brown beneath. It is of somewhat broad-shouldered and stocky stature.

Come February and the oriental skylark begins its astonishing performance. It climbs steeply into the sky and, fluttering moth-like in wavy circles, pours forth a stream of melody for longer than you can look up without wincing. At last, after five or ten minutes, the little bird falls silent and drops to earth, vanishing uncannily into the ground. And then within seconds, has launched itself out of the grass tufts for an encore. There are usually four or five skylarks displaying thus, over a single field or area, and occasionally they are so high up that you only hear the music. It seems as though some magical musical box has been switched on in the heavens.

Purple Sunbird, *Nectarinia asiatica*, Shakkar Khora, Phool Soongni

This tiny (10 cm), iridescent sunbird, clad in metallic midnight blue and purple has probably got more city people excited about, and interested in, birds than any other species. A common high-voltage visitor to every garden and park in Delhi, it probes blooms for nectar with its long, curving, slender bill, hovering briefly before the bloom and immediately causing its admirers to exclaim, 'Hummingbird!'

Only the male in breeding plumage wears (between February and August) the glamorous midnight blue and purple outfit—embellished, no less, with scarlet and yellow tufts under the armpits! The female is a dull, nondescript olive brown and yellowish bird, and in the non-breeding season (August to February) the males are similarly dressed, but wear a black or purple necktie as mark of identification.

By January, and certainly February, the males change into their glittering costumes and zip after each other like huge, shrill carpenter bees, as they try and establish territories. They call out sharply and excitably, 'Wich-wich-wich!', as they blur after one another, pausing now and then to sing their high-spirited 'Cheewit-cheewit-cheewit!' song, which can serve as a mood elevator no matter what species you are.

Females build oblong purse-like nests out of string, grass, and straw, bound together with cobweb and camouflaged to look like a bag of rubbish. Two to three brown-marked greyish or greenish-white eggs are laid and both parents look after the brood.

Sunbirds sustain their high-voltage life-styles on a diet of nectar, small insects, and succulent spiders. They are important pollinators.

House Sparrow, *Passer domesticus*, Gauraiya, Churi

Easily the capital's commonest (and most noisy) small bird, the house sparrow (15 cm) is happy to share its living space with us, nesting unperturbed in our balconies, garages, ceiling fans, and any suitable hole in the wall or gap in the bookshelf that catches its fancy.

The cock sparrow is quite a handsome rake with his black bib, cigarette-ash crown, and black-streaked chestnut plumage. The hen is more ordinary and streaky looking in light and dark brown. House sparrows are omnivorous and will eat anything from *dana* to bread crumbs and kitchen leftovers.

In many of Delhi's parks, gardens, and leafy colonies, the new day is welcomed by the shrill, chattering, dawn chorus of hundreds of house sparrows as they 'wash and dress up' and prepare to fly down to the various feeding spots where, they know, kind-hearted souls (or those seeking absolution from hideous sins) will have laid out breakfast.

House sparrows nest between January and October, raising as many families as they can during this period. Straw, grass, and rubbish are stuffed into a hole in a wall, the ceiling or bush—or gap on top of the ceiling fan so that they can dice with death every time they enter or leave their homes. Three to five pale greenish-white eggs are laid and both parents bring up the brood.

The cock sparrow does a pompous little courting routine, sticking out his chest, drooping his wings, arching his rump, and strutting before his beloved to woo her. When pairs nest in close proximity, horrendous soap-opera-type melodramas can take place, with gross infidelities suspected all round, and vendettas carried out far beyond the call of duty. Common though they may be, house sparrows are well worth watching.

Chestnut-shouldered Petronia (Yellow-throated Sparrow), *Petronia xanthocollis*, Jungli Chiria

Slightly smaller than the House Sparrow, the chestnut-shouldered petronia (13.5 cm) is a dusky-brown sparrow, with two white wing bars and a chestnut shoulder patch. Males have a smudge of pollen yellow on the breast. The hens are similar but more faded, and do not have the yellow patch. During the breeding season the cocks have black bills, the females pinkish ones. Males sing rather like house sparrows but more mellifluously 'Chillup chillup chillup!' as the lyrics go.

Chestnut-shouldered petronias like thorny scrub country and open dry forest. Two areas where I have always seen them are the rocky scrublands arounds the Badkhal Lake complex and at Buddha Jayanti Park—where you can get close to them as they come down to partake of the handouts.

White-browed Wagtail (Large Pied Wagtail), *Motacillia maderaspatensis*, Mamula

A large (21 cm), somewhat elongated-looking black and white wagtail that is resident and lives alongside the river and other suitable water-bodies such as Badkhal Lake. The head is black (with a prominent white eyebrow), as is the back and breast. There is white on the wings; the belly, abdomen, and tail are white too.

Nearly always seen in pairs, white-browed wagtails nest (between March and October) in holes in walls or trees, or under bridges and culverts, always near water. They build cup-shaped structures out of grass and rootlets. Three or four greyish, brownish, or greenish-white eggs are laid, marked with brown.

The wagtail has a distinct 'Chiz-zat!' call and a magpie-robin-like whistling song. The bird is insectivorous.

Migratory Wagtails

These can cause some confusion as of the four species visiting Delhi, three are clad in various combinations of olive yellow, ashy grey, and sulphur. Worse, there are several races within two of these three species and, complicating matters even further, they are not suited out in their distinctive breeding best while visiting the capital, but are in pale, indistinct grey and olive casuals!

The **White Wagtail** (*Motacilla alba*, Dhobin) at about 21.5 cm is thankfully distinctive in its smart ash-grey, black and white plumage. Ash grey on top, it has a black patch on the back of its head and a black bib. The forehead, sides of the head and underside are white; the wings black and white.

White wagtails trundle about on lawns, and in parks, compounds, car parks, roadsides and nearly everywhere in Delhi in winter. They arrive by around September and stay on till April. They are somewhat rotund and jaunty birds, bobbing their heads and wagging their tails as they go about scouting for insects. They fly in the typical dipping manner of their species.

The **Citrine Wagtail**, formerly the Yellow-headed Wagtail (*Motacilla citreola*, Pani-ka-pilkiya) can occasionally be spotted with its trademark bright yellow head (especially towards April) and underparts, which make it easy to identify. The upper parts and rump are usually grey, or may even be black depending on the race. In drab, typical mid-winter uniform, only smudges of yellow are discernible about the face, the body being ashy above and off-white below.

Citrine wagtails play hide and seek in damp, squelchy fields, usually near water, or hop about amongst reeds or water hyacinth in the river or water-bodies such as Sultanpur and Badkhal Lake.

The **Yellow Wagtail** (*Motacilla flava*, Pilkiya) 18 cm, comes in as many as four races (depending on the colour of the head!), but which, while visiting Delhi in winter, are usually olive greenish brown above, with brighter yellow breasts. They arrive by around September and occupy damp agricultural fields, staying on till April and perhaps early May.

The **Grey Wagtail** (*Motacilla cinerea*) 20 cm, is blue grey above and off white underneath, suffusing to sunshine yellow towards the tail. It has a white supercillium. It is thought to be a passage migrant, passing through Delhi between September and October and then again, in March and April.

Paddyfield Pipit (Indian Pipit), *Anthus rufulus*, Rugail

Pipits look rather like outsized hen sparrows, but carry themselves more upright and don't hop furtively but canter and run. The paddyfield pipit is a slim, somewhat long-tailed bird, streaked dark and light brown above, pale bland mushroom on the belly, but with more streaking on the breast. It has a white eyebrow (supercillium) and the feathers of the rump are dark brown outlined with fulvous.

Paddyfield pipits run about in areas of dry stubble, short grass, sand, and muddy tracts near water—such as the river banks and the areas around Sultanpur jheel—as well as in cultivated fields and wastelands. They are ground nesters (nesting occurs between March and June), constructing a shallow cup of grass and rootlets in a hardened hoof print or under a clod. Three to four yellowish- or greyish-white eggs marked with brown are laid. Both parents look after the brood. Pipits are insectivorous. Their flight is undulating, usually accompanied by a high-pitched 'Tsip tsip!' or 'Pipit pipit!' squeaking.

The Weavers

Easily the best known of the weavers is the **Baya Weaver Bird** (*Ploceus philippinus*, Baya) whose elegant vase-like nest, woven expertly out of plaint grass and stems hangs from branches usually in trees overlooking water. Rather sparrow-like in size (16 cm) and stature, but with a more finch-like bill, the males in the breeding season (May to September) have a vivid pollen-yellow head, breast and belly, offset by a dark chocolate face. The fulvous upper parts are also flaked darkly with chocolate and appear dusted with pollen around the shoulders. Females and males in non-breeding plumage are rather nondescript hen-sparrow-like birds.

Bayas are fairly common in the woodland areas around Delhi, especially if there is a water body in the vicinity. You can, for instance, often meet a large flock at Buddha Jayanti Park adjoining the Ridge, indulging in power breakfasts with silverbills and sparrows. Another colony used to nest alongside a filthy, furtive pond on the hilly section of the northern Ridge opposite Hindu Rao Hospital. The date palms on the island of the main duck pond of Delhi zoo are usually also festooned with baya nests, as are those studding the gardens around Humayun's tomb. The agricultural country south of Okhla Barrage also supports colonies in suitable niches, often alongside some rain-filled ditch or nallah.

The shrill excited 'Chee-chee-chee!' calls of scores of breeding males inviting females to their half-built homes carry a long way and are a prominent indicator of the presence of the birds.

Another common member of this clan is the **Streaked Weaver** (*Ploceus manyar*, Bamani Baya). During the breeding season—from May to September—the males have fog-lamp golden crowns and dark burnt-wood faces, slightly ashy at the chin. The upper parts are flaked in dark and light brown, both flakes and dark ground colours paling increasingly and fading out on the breast and belly. During the non-breeding season they look like hen sparrows, with turmeric-stained supercilliums and a pale transverse streak curving up from the cheeks.

Streaked weavers can be met along the reed-fringed edges of the river, the scrub forests of the Ridge, and similar habitats. Their nests, also woven out of plaint grass are truncated editions of the bayas' nests, with shorter entrance tunnels and more firmly fixed to bulrushes and reeds so that they don't swing freely. Streaked weavers have been nesting along the reed-fringed edges of the moats at Delhi zoo—and, on one occasion, a pair with

Z+ category security, built their home in the tiger's enclosure. For many years, a colony has been trying in vain to nest beside the ponds of the northern Ridge, but the appalling decibel levels, (caused by lunatic walkers and exercise fanatics—and, recently a hooligan flock of domestic geese) have put paid to their efforts. Truth be told, the streaked weavers too make a hell of a screeching racket at this time—which can quite get on your nerves.

The **Black-breasted Weaver** (*Ploceus benghalensis*, Sarbo Baya) 14 cm, is quite similar to the streaked weaver, golden capped like it, but with a dark burnt brown breast, paling towards the chin, with no streaking. Its nest too is similar to that of the streaked weaver. Nowhere as common as the other two, they can be met along the reedy banks of the Jamuna and grassland country south of Okhla. It pays to look carefully at every apparent streaked weaver, just in case.

Weavers eat seeds and insects. They may be outrageously bigamous, and a male baya, having settled one wife snugly in a beautiful home may go about constructing another for a second one and so on. Two to four white eggs are usually laid.

Avadavats and Munias

Be warned: these sparrow-sized attractive and charming birds are big favourites with the pet trade and often dyed golden, metallic green, and maroon to make them appear more exotic. In its natural state, the **Red Avadavat** or **Red Munia** (*Amandava amandava*, Lal Munia) 11 cm, is probably the most striking of the lot. During the breeding season—between June and September—the male is clad entirely in crimson, dusted with white on the sides of its neck and body. A broad, dark crimson-black line dramatizes the area between the eye and the conical crimson bill. During the non-breeding season, the males resemble their wives—dusky brown on top, darkening at the wings (dotted with white) and crimson on the rump, which makes them look as though they've been shot in the bottom. The throat, breast, and underside are pale beige.

While not too common, and always a joy to behold, some strongholds of the red avadavat are the Jawaharlal Nehru University campus (especially during the monsoons), the dusty paths leading off from the river, and areas where grass scrub and water are in close proximity. I have seen nest building in progress at Sultanpur and once on the northern Ridge, where the foolish bird, instead of choosing the path less trodden, chose one that was too heavily trodden and had to give up the attempt. A tennis-ball-like cup of grass is made low down in the reeds if the attempt is successful.

The **Silverbill** or **White-throated Munia** (*Lonchura malabarica*, Charchara/Charga/Charakka/Pitta) in its mushroom beige and butterscotch brown is certainly Delhi's commonest munia. Its flight feathers are black as is its sharp pointy tail. The conical bill is silver blue.

Silverbills love the scrub country of the Ridge, Buddha Jayanti Park, the gardens around Humayun's Tomb, and a flock frequents the garden next door every morning where breakfast is laid out for them. The birds are known to nest throughout the year and, here in Delhi, the monsoons seem to be a better time than most. I have seen them nesting at Sultanpur—an untidy globular structure of coarse grass was placed in the most viciously armed acacia bush—and made me wonder how on earth the adults flew in and out without impaling themselves and what the fledglings would do when confronted with their first flight. Four to six pure white eggs are laid and both parents bring up the brood. Silverbills have a pleasant chirruping call.

The handsome **Scaly-breasted Munia** (*Lonchura punctulata*, Telia Munia) 15 cm, is an upright-standing munia with shimmering chestnut to wine-red upper parts, darkening around the face, and a white breast and belly, boldly scalloped in black, which makes it look as though it were clad in chain mail. The rounded tail is brownish.

Not as common as the previous two, the scaly-breasted munia can sometimes be met with in grassland country and parks. I have seen them (nesting) at Sultanpur, on the northern Ridge and at

University Gardens, where again the specimen concerned appeared to be engaged in nest building. The nesting season lasts from July to October, and a globular nests is constructed out of grass and placed low down in a bush. Four to eight white eggs are laid and both parents look after the chicks.

The **Black-headed Munia** (*Lonchura malacca*, Nakal Nar) 14 cm, is a chestnut and black munia, of which one race has a white belly and undersides and a black vent, as compared with the other's entirely black and chestnut plumage. I have seen the black and chestnut version on a few occasions in the reed beds along the Jamuna, and fields south of Okhla, but the birds are certainly not as common as the other members of their family.

Commom Rosefinch, *Carpodacus erythrinus*, Lal Tuti

This fluffed-up-looking finch with a stout conical bill looks as though it has been helping itself to a jar of strawberry jam. Its head, chin, and throat are deep strawberry pink, the upper parts beige brown tinged with pink (the jam gets everywhere!), the pale off-white belly streaked pinkish, further reiterating the streaked strawberry effect. The females do not dip into the jam and are olive brown, with fine streaking on the head, breast and sides of the throat, and have two white wing bars.

Rosefinches are winter visitors and stay in small numbers between Septembr and April. The reed-fringed areas along the river and surrounding grasslands are favourite haunts as are the scrub woodlands of the Ridge. They may join silverbills and sparrows at the feeding spots in parks (Buddha Jyanti Park, for example) and so it is always sensible to check these out. The number of birds increases towards April when additions stop over en route to their breeding grounds up north (in the Himalayas). I have seen them on the northern Ridge and at Deer Park (helping themselves to mulberries not strawberries!) and Buddha Jayanti Park.

Index of Birds Featured in the Rogues' Gallery

~~~~

Alexandrine Parakeet (Large
    Indian Parakeet) 63
Asian Koel 61
Asian Openbill (Open-billed
    Stork) 109
Asian Paradise Flycatcher 115
Asian Pied Starling (Pied Myna)
    124
Avadavats 144

Babblers 133
Bank Myna 125
Barheaded Goose 40
Barn Owl 66
Bitterns 104
Black Drongo 115
Black Francolin (Black Partridge)
    38
Black Kite (Pariah Kite) 91
Black Redstart 120
Black-bellied Tern 90
Black-crowned Night Heron
    (Night Heron) 104
Black-necked Stork 109
Black-rumped Flameback (Lesser
    Golden-backed Woodpecker)
    50

Black-shouldered Kite (Black-
    winged Kite) 90
Black-tailed Godwit 75
Black-winged Stilt 80
Blue-capped Rock Thrush 116
Blue-cheeked Bee-eater 59
Bluethroat 118
Blyth's Reed Warbler 131
Brahminy Starling (Black-headed
    Myna) 122
Bronze-winged Jacana 82
Brown-headed and Black-headed
    Gulls 87
Brown-headed Barbet (Green
    Barbet) 51
Bulbuls 128

Chestnut-shouldered Petronia
    (Yellow-throated Sparrow)
    140
Chiffchaff 133
Clamorous Reed Warbler
    (Indian Great Reed Warbler)
    131
Comb Duck (Nakta) 42
Commom Rosefinch 146
Common (or Fantail) Snipe 74

Index of Birds

Common Coot (Coot) 74
Common Crane 71
Common Hawk-cuckoo or Brainfever Bird 60
Common Hoopoe (Hoopoe) 54
Common Kestrel (Kestrel) 97
Common Kingfisher (Small Blue Kingfisher) 56
Common Moorhen (Indian Moorhen) 74
Common Pochard 48
Common Redshank (Redshank) 75
Common Sandpiper 77
Common Starling 123
Common Stonechat (Collared Bushchat) 120
Common Teal 45
Coppersmith Barbet (Coppersmith) 52
Cormorants 99
Cotton Pygmy Goose (Cotton Teal) 43
Crested Serpent Eagle 93

Darter (Indian Darter) 98
Demoiselle Crane 72

Egrets 100
Egyptian Vulture (Scavenger Vulture) 92
Eurasian Collared Dove (Indian Ring Dove) 68
Eurasian Golden Oriole (Golden Oriole) 113
Eurasian Marsh Harrier (Marsh Harrier) 94
Eurasian Spoonbill (Spoonbill) 107
Eurasian Thick-knee (Stone Curlew) 79
Eurasian Wigeon (Wigeon) 44

Gadwall 43
Gargeny (Blue-winged Teal) 46
Great White Pelican (Rosy Pelican) 107
Greater Adjutant (Adjutant) 110
Greater Coucal (Crow Pheasant) 62
Greater Flamingo 105
Greater Painted Snipe (Painted Snipe) 79
Green Bee-eater (Small Green Bee-eater) 58
Green Sandpiper 76
Greenish Warbler (Greenish Leaf Warbler) 133
Grey Francolin (Grey Partridge) 37
Grey Heron 103
Grey-headed Canary Flycatcher 117
Greylag Goose 40
Gull-billed Tern 89

House Crow 112
House Sparrow 139
House Swift (Little Swift) 65

Ibises 106
Indian Chat (Brown Rock Chat) 122
Indian Grey Hornbill (Common Grey Hornbill) 53
Indian Myna 124
Indian Peafowl (Common Peafowl) 39
Indian Pond Heron 102
Indian Robin 119
Indian Roller or Blue Jay 55

Large-billed Crow (Jungle Crow) 113
Larks 136

Index of Birds ◆ 149

Laughing Dove (Little Brown Dove) 68
Lesser Whistling Duck (Lesser Whistling Teal) 41
Lesser Whitethroat 132
Little Grebe (or Dabchick) 97
Little Heron (Little Green Heron) 103
Little Ringed Plover 83
Little Stint 78
Long-legged Buzzard 96

Mallard 44
Marsh Sandpiper 76
Martins 126
Munias 144

Northern Lapwing (Lapwing or Green Lapwing) 86
Northern Pintail (Pintail) 46
Northern Shoveller (Shoveller) 47

Oriental Honey Buzzard (Honey Buzzard) 95
Oriental Magpie Robin (Magpie Robin) 118
Oriental White-eye 130

Paddyfield Pipit (Indian Pipit) 142
Painted Stork 108
Pheasant-tailed Jacana 82
Pied Avocet (Avocet) 81
Pied Bushchat 121
Pied Cuckoo (Pied Crested Cuckoo) 59
Pied Kingfisher 57
Plum-headed Parakeet (Blossom-headed Parakeet) 65
Prinias 129
Purple Heron 102
Purple Sunbird 138

Purple Swamphen (Purple Moorhen) 73

Red Collared Dove (Red Turtle Dove) 69
Red-crested Pochard 48
Red-throated Flycatcher 117
Red-wattled Lapwing 84
River Lapwing (Spur-winged Plover) 86
River Tern (Indian River Tern) 89
Rock Pigeon (Blue Rock Pigeon) 67
Rose-ringed Parakeet 64
Rosy Starling (Rosy Pastor) 123
Ruddy Shelduck (Brahminy Duck) 42
Ruff (and Reeve) 78
Rufous Tree Pie (Indian Tree Pie) 112

Sarus Crane 70
Shikra 94
Shrikes 111
Small Minivet 114
Small Pratincole (Small Indian Pratincole) 87
Spotbilled Duck 45
Spotted Owlet 67
Swallows 126

Tailor Bird 132
Tawny Eagle 96
Tufted Duck (Tufted Pochard) 48

Wagtails 141
Weavers 142
White-breasted Waterhen 72
Whitebrowed Fantail 114
White-browed Wagtail (Large Pied Wagtail) 140
White-rumped Vulture (White-backed or Bengal Vulture) 92

White-tailed Lapwing 85
White-throated Kingfisher
  (White-breasted Kingfisher)
  56
Wood Sandpiper 77
Woolly-necked Stork (White-
  necked Stork) 109

Yellow-crowned Woodpecker
  (Yellow-fronted Pied Wood-
  pecker/Marhatta Woodpecker)
  49
Yellow-footed Green Pigeon
  (Green Pigeon) 70
Yellow-wattled Lapwing 83